Astronomers' Universe

Other titles in this series

Rejuvenating the Sun and Avoiding Other Global Catastrophes
Martin Beech

Origins: How the Planets, Stars, Galaxies, and the Universe Began
Steve Eales

Calibrating the Cosmos: How Cosmology Explains Our Big Bang Universe
Frank Levin

The Future of the Universe
A.J. Meadows

Dr. Lucy Rogers

It's ONLY Rocket Science

An Introduction in Plain English

Dr. Lucy Rogers CEng MIMechE FRAS
Isle of Wight, UK.
www.itsonlyrocketscience.com

ISBN 978-0-387-75377-5 e-ISBN 978-0-387-75378-2
DOI: 10.1007/978-0-387-75378-2

Library of Congress Control Number: 2007939660

9 8 7 6 5 4 3 2 1

Springer Science + Business Media

springer.com

For Laura and Hannah

Acknowledgements

There are many people and organizations that have helped, either directly or indirectly, to make this book a reality. I would first like to thank the British Association for the Advancement of Science and the Guardian Newspaper, for encouraging me, as an engineer, to become involved with the media. I was fortunate enough to be awarded one of the BA's Media Fellowships at the Guardian newspaper, and this scheme opened my eyes to the possibility of sharing science with everyone, and not just limiting it to academia and industry. Without this scheme and the wise words of Tim Radford, the then science editor at the Guardian, I would never have started writing.

I would also like to thank John Thomson and West Didsbury Astronomical Society for rekindling my interest in astronomy, and to Starchaser Industries for the opportunity to actually play, I mean work, with rockets through all stages of development.

My thanks also go to the team at Springer, particularly John Watson, who first believed in the project, and also to Harry Blom, my editor, and his assistant editor Chris Coughlin.

I have been constantly amazed at the generosity of the many people who have taken the time to answer my questions and explain facets of rocket science that, at the start of the project, I did not even know were involved. These include the staff at NASA, in particular Kylie Clem, Allard Beutel and Jennifer Ross-Nazzal and also to Colin R. McInnes, A.D. King, Russell Eberst, Claude Phipps, Gregory Benford and also the members of the HearSat email list.

Neil Chance read the first draft of each chapter and, with queries and probing questions, made me think much more deeply than I had originally intended. The book is much clearer because of this, and I thank him for it. I would also like to thank Jan Foy and John Langley for their encouragement and time.

I must also thank my brother, Benjamin Rogers, who, since our school days, has explained many aspects of science and maths clearly and patiently. He has provided invaluable support when my understanding of the physics involved in rocket science wavered. However, any mistakes in the book are mine.

Finally my love and thanks go to my parents and to my partner Stephen J. Griffiths, who have supported and encouraged me throughout the whole process of preparing this book.

Contents

1. Introduction

Non est ad astra mollis e terris via
(There is no easy way from the Earth to the stars)
Seneca, circa AD 50

On October 4, 1957, *Sputnik 1* became the first artificial satellite. It was launched into orbit by the former Soviet Union. The media coverage following the Soviet's success meant that the general public quickly became aware that rocket science was a scientific endeavour and no longer in the realms of science fiction. Rocket science has always been perceived as very challenging and the difficulties the Americans faced with their early launch failures reinforced this idea. Wernher von Braun, a major contributor to the development of rocket technology, both in Germany and later in the USA, said:

> It takes sixty-five thousand errors before you are qualified to make a rocket.

After the success of *Sputnik 1*, the launch and operation of satellites became very politically sensitive and so the brightest scientists and engineers were often employed as rocket scientists. It therefore became thought of as a subject only for the most intelligent. There are other fields of study that are arguably more challenging than rocket science, but, other than brain surgery, none have entered the mainstream vocabulary as a difficult thing to do.

This book aims to explain, in everyday terms, just what is involved in launching something into space and exploring the universe outside of our own small planet. It provides an overview into what is required for a mission, without the mathematical analysis of the fine detail. Such analysis is included in many good textbooks, some of which are listed in the bibliography. The rest of this chapter explains and defines some of the fundamental properties of space and rocket science that will be referred to throughout the book. The more technical aspects have been relegated to the Appendices, and, for simplicity, I have usually referred to all spacefaring humans as astronauts, no matter their citizenship or the country from which they launched.

Space

There is no clear boundary between the Earth's atmosphere and space. The molecules of air just become further apart the higher up you go. In the middle of the 1950s a definition for the boundary between the atmosphere and space, known as the Kármán line, was fixed at an altitude of 100 kilometres. The Federation Aeronautique Internationale (FAI), the world air sports federation, established this line. The FAI is a non-governmental and non-profit making organization with the basic aim of furthering aeronautical and astronautical activities worldwide. The Kármán line is still used and is internationally accepted as the boundary to space for the purposes of world records and many treaties. However, the USA uses an altitude of 80 kilometres for awarding astronaut's wings. Noctilucent clouds, the electric-blue cloud formations that are thought to be composed of small ice-coated particles, form at an altitude of about 80 kilometres and this is also about as high as was reached by plumes of ash hurled by the eruption of Mount Krakatoa in Indonesia in 1883. At the Kármán line boundary there is still a perceptible drag caused by the atmosphere. During re-entry, when a spacecraft returns from space to the Earth, this drag becomes noticeable at about 120 kilometres and the effect of friction, due to the atmospheric particles, becomes evident as heat. This 120 kilometres boundary is called the entry interface. Table 1.1 compares different objects and their height above the Earth.

Solar Wind and the Van Allen Radiation Belts

The solar wind is a constant stream of particles that leaves the Sun. These particles include ions from almost every element in the periodic table, but the majority are electrically charged subatomic particles called protons and electrons. The number of particles making up the solar wind varies with the amount of activity on the surface of the Sun.

The solar wind extends to form a bubble around our solar system called the heliosphere. The blurred edge where the heliosphere

Table 1.1 Heights Above the Earth of Different Objects

Height above sea level (km)	Object
8.8	Mount Everest
9	Most passenger aircraft
30	Military jets
80	The USA awards Astronaut's Wings to those who fly to this altitude
80	Mount Krakatoa's ash plume in 1883
80	Noctilucent Clouds
100	Kármán line – boundary between atmosphere and space for the purposes of world records and many treaties
120	Entry interface – height at which friction, due to atmospheric particles, starts to heat spacecraft re-entering the Earth's atmosphere
390	International Space Station
35,880	Satellites in geosynchronous orbit, such as those that broadcast satellite TV
385,000	The distance to the Moon
150,000,000	The distance to the Sun

meets the interstellar gas outside of our solar system is called the heliopause. Nothing man-made has yet reached the heliopause, although NASA's space probes *Voyager 1* and *2* and *Pioneer 10* and *11* are nearing it.

Most of the particles that come near to the Earth are deflected by the Earth's magnetic field. However, some become trapped by the magnetic field and stay in one of two belts, known as the Van Allen belts. These look similar to two huge doughnuts or car tyres encircling the Earth and the only parts of the Earth not covered by the belts are the areas surrounding the north and south poles, as can be seen in Figure 1.1 overleaf.

Scientists predicted that these belts would exist but their presence was not confirmed until 1958 when an instrument on the satellite *Explorer 1* detected them. The belts are named after the leader of the team conducting the satellite experiment, Dr. James Van Allen. The radiation caused by the particles in the belts can cause problems to satellites, as it can degrade components, particularly semiconductor and optical devices, induce background noise in detectors and cause errors in digital circuits. It is also a threat to astronauts, as is discussed in Chapter 8 – "Humans in Space".

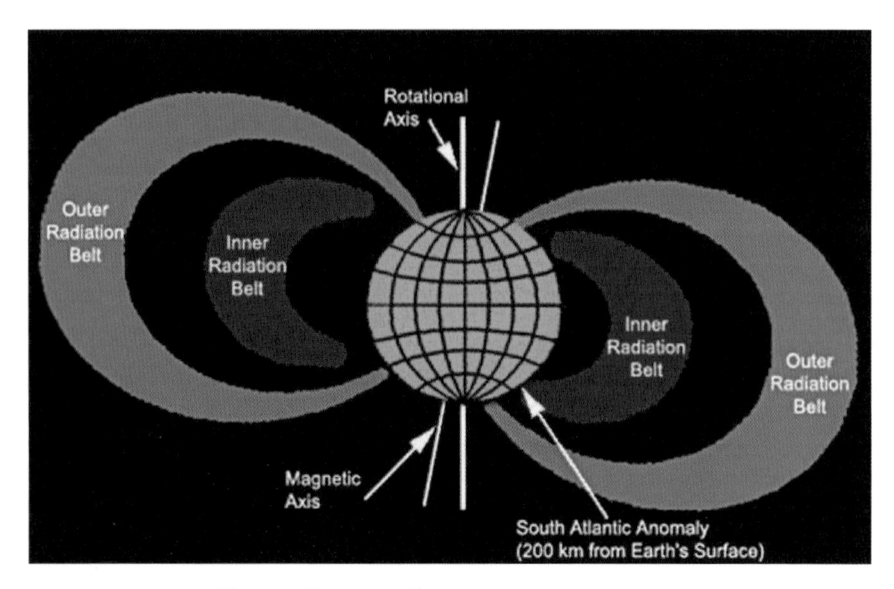

FIGURE 1.1 Van Allen Radiation Belts.
Image courtesy NASA

One belt is a lot lower than the other and is called the lower or inner belt. It starts at about 1,000 kilometres above the Earth's surface, depending on the latitude, and extends to about 10,000 kilometres. This belt contains mainly positively charged protons. The region around 3,500 kilometres has the highest density of protons and it is therefore this region that produces the most damaging radiation effects. Part of the inner belt dips down to about 200 kilometres above sea level in a region over the southern Atlantic Ocean, off the coast of Brazil. This is caused by two main factors. First, the Earth's magnetic axis and rotational axis are not perfectly aligned, which gives rise to the difference between true north and magnetic north, and second, the centre of the magnetic field is not at the geographical centre of the Earth. The dip in the belt is known as the South Atlantic Anomaly and special precautions are taken when spacecraft pass through this region.

The higher belt is known as the outer belt and consists mainly of negatively charged high-energy electrons. It starts at about 10,000 kilometres above the surface of the Earth and reaches up to about 65,000 kilometres, but the region around 16,000 kilometres causes the most radiation damage to satellites and living things. As the two Van Allen belts cause problems to spacecraft they are generally avoided by limiting the orbit to below or above the main areas of radiation.

Although most of the solar wind is either diverted by the Earth's magnetic field or trapped in the Van Allen Belts, some solar particles reach the top of the atmosphere and collide with the molecules of air. This collision causes the air molecules to heat up and glow. If it is dark enough on the surface of the Earth, the light caused by these glowing air molecules is visible from the ground. In the northern hemisphere these lights are known as the aurora borealis or northern lights and in the southern hemisphere as the aurora australis.

Gravity

Gravity causes all objects to attract each other. On the Earth the effect of gravity can be seen when an item, such as an apple, falls to the ground. The Earth and the apple are attracted to each other with the same-sized force. However, as the apple is much lighter than the Earth it accelerates much faster, until they crash into each other. The apple usually comes off worse.

Gravity acts along a line between the two centres of the items involved, in the above example this is between the centre of the apple and the centre of the Earth. Therefore, gravity will ensure that, on the Earth, objects are always attracted towards the centre of the Earth or, as seen from the surface, downwards. The Earth will always be attracted towards the centre of the object, but this is not noticeable to us.

The size of the force depends on the mass of the objects and the distance between them. A larger mass produces a greater force. Weight is a measure of the force of gravity acting on a mass. With less gravity, an item will weigh less but it will still have the same mass. Just as the mass of the Earth produces a gravitational attraction, so the other planets and the Sun also produce a gravitational attraction relative to their masses.

The size of the force of attraction from an object decreases rapidly with distance. If an apple were twice as far away from the centre of the Earth, about 6,380 kilometres above the surface of the Earth, it would only "feel" a gravitational attraction of a quarter the size of what it "felt" when on the surface of the Earth. It would therefore accelerate towards the Earth more slowly and it would only have a quarter of the acceleration it had on the surface. This "twice the distance, quarter of the size" rule is called the inverse square law and applies to many other things, including the amount of light received from a source. For example, if a star were twice as far away, it would appear to be only a quarter as bright. The English physicist Sir Isaac Newton first published the equations describing gravity in 1687. The story goes that Newton was sitting under an apple tree when an apple fell on his head. From that, he worked out that gravity on the Earth and gravity in space are the same. However, it is probably more likely that Newton was thinking about the Moon falling to the Earth and comparing it to an apple. He realised that the force of gravity must get weaker further away from the Earth.

To overcome gravity, energy is required. To throw a ball upwards requires effort. The harder a ball is thrown, the further it will go before slowing down and falling back to Earth. The effort is used to overcome the attraction due to gravity as it moves upwards. If you keep throwing the ball harder there will come a time when it would continue moving away from the Earth, slowing down all the time. It would get slower and slower and never quite stop. If you could wait forever it would eventually stop, but by then it would have travelled an infinite distance. This velocity is called the escape velocity. From the surface of the Earth this value is about 11.2 kilometres per second or about 40,320 kilometres per hour, however, the value decreases with distance away from the surface. When a space vehicle approaches another body, such as the Moon or a planet, there comes a time when the gravity from that body attracts it more than the Earth attracts it. At this stage, effort is no longer required to escape from the Earth, but effort would now be required to get back to the Earth. If the spacecraft is travelling slow enough, it will be captured by the other body and enter into orbit around it. If the spacecraft was travelling too fast to be captured its trajectory would be bent by the object, and the spacecraft would fly by the object and off into space. This type of fly-by is also known as a gravity assist, and is described in more detail in Chapter 4 – "Movement in Three Dimensions".

It is possible to overcome gravity and travel in any direction in space. However, this requires continual thrust from a rocket, which requires a lot of power. With present technologies, use of this amount of power and propellant is not practical and so all motion in space is essentially coasting around celestial bodies that have a large gravitational force. Alterations to the orbits are made by relatively short bursts of thrust from rocket motors. The effect of gravity is almost certainly going to remain an important consideration in spaceflight, even with better technologies. Some possible new technologies are described in Chapter 11 – "The Future".

Propulsion

On the Earth, forward motion is usually achieved by pushing on some medium, such as the ground for a car and the sea for a motorboat. We walk forwards by pushing back against the floor with our feet. This is why it is difficult to walk on ice. Although most propulsion systems do push on something, the act of throwing something out in the opposite direction can also produce forward motion. This can be seen if a child carries a heavy ball while standing on a skateboard. If the child throws the ball away, both the child and the skateboard will move in opposite directions. A jet aeroplane works in a similar way, it takes in air, squeezes it in a compressor, mixes it with fuel, and the gases from the resulting explosion are thrown out behind at a faster speed. These methods of propulsion use an action that causes a reaction, which was described in 1687 by Sir Isaac Newton in his third law of motion "For every action there is an equal and opposite reaction". Although Newton's third law of motion about action and reaction sounds simple, it is often misunderstood and many people assume that for something to be propelled, it must push against something.

Jet propulsion is any form of reaction motor that ejects matter, and is therefore also known as a reaction motor. Whatever is ejected is called the propellant. High up in the atmosphere the molecules of air are far apart, and at altitudes greater than about 80 kilometres the atmosphere no longer exists as an effective medium as it is almost a vacuum. A jet aeroplane therefore cannot work at these altitudes as it requires the oxygen in the air to act as an oxidiser, which enables the fuel to burn. A reaction motor that carries all of its propellant and, if required, the fuel and oxidiser with it, can overcome an absence of air. A toy balloon filled with air is a reaction motor, although one that only requires propellant and no fuel. When the neck is released, the balloon flies around the room as the air is expelled until it is empty, when it falls to the ground. This would happen even if there were no air in the room. Rockets also work in the same way. They carry all of their propellant with them and as it is pushed out of the rocket, usually as a gas, the rocket is pushed forwards.

In 1919, Robert H. Goddard, the American physicist and early rocket pioneer, published a treatise entitled *A Method of Reaching Extreme Altitudes.* This technical paper outlined his ideas on rocketry and also included calculations and results from various tests he had carried out. However, some newspaper editors had misunderstood Newton's third law. The editorial on page 12 of *The New York Times*, January 13, 1920, dismissed the idea of a rocket travelling in a vacuum and ridiculed Goddard by saying:

> After the rocket quits our air and really starts on its longer journey [towards the Moon], its flight would be neither accelerated nor maintained by the explosion of the charges it then might have left. To claim that it would be is to deny a fundamental law of dynamics only Dr Einstein and his chosen dozen, so few and fit, are licensed to do that.

It continues:

> That Professor Goddard, with his 'chair' in Clark College and the countenancing of the Smithsonian Institution, does not know the relation of action to reaction, and of the need to have something better than a vacuum against which to react – to say that would be absurd. Of course he only seems to lack the knowledge ladled out daily in high schools.

In the same article, *The New York Times* also dismissed Jules Verne, the 19th-century French science fiction author, saying he:

> deliberately seemed to make the same mistake that Professor Goddard seems to make.

It continues:

> That was one of Verne's few scientific slips, or else it was a deliberate step aside from scientific accuracy, pardonable enough in him as a romancer, but its like is not so easily explained when made by a savant who isn't writing a novel of adventure.

This, and other, inaccurate reporting encouraged the public's perception that rocket propulsion and space travel were impossible, and also that Goddard himself was wasting time and money investigating such things. It took almost 50 years for the newspaper to publish a correction. This was done on page 43 of *The New York Times*, July 17, 1969, by which time the *Apollo 11* crew were well on their way to the Moon:

> A Correction. On Jan. 13, 1920, 'Topics of the Times' an editorial-page feature of The New York Times, dismissed the notion that a rocket could function in vacuum and commented on the ideas of Robert H. Goddard, the rocket pioneer.

It continues:

> Further investigation and experimentation have confirmed the findings of Isaac Newton in the 17th century, and it is now definitely established that a rocket can function in a vacuum as well as in an atmosphere. The Times regrets the error.

The lack of atmosphere is one of three main differences between travelling within the Earth's atmosphere and outside of it. This can be a problem in as much as there is no available oxygen to help burn the fuel, but it can also be an advantage, as there is no air resistance to slow the craft down. The other differences are the vast distances involved and both the changes in the force of gravity with distance from the Earth, and also the influence of the gravity from the Moon, Sun and planets. Because of these differences, and the difficulties involved overcoming them, the term astronautics has been applied to cover "The art or science of locomotion outside the Earth's atmosphere" as defined by the Oxford English Dictionary.

The distances involved travelling outside the Earth's atmosphere are much larger than travelling within the atmosphere. The maximum distance to get from any point to any other point on the Earth, travelling on, or near, the surface, is about 20,000 kilometres. The distance from the Earth to the Moon is about 385,000 kilometres, which is almost 20 times further, and the distance from the Earth to the Sun is about 150 million kilometres, or 7,500 times farther than any two places on the Earth. Sunlight takes over eight minutes to travel from the Sun to the Earth and, so far, humans have only managed to travel at a small fraction of the speed of light. Travelling as fast as a fighter jet, about Mach 1.5 or 1,800 kilometres per hour, it would take about nine and a half years to reach the Sun. If we travelled in a straight line at 150 kilometres per hour, which is about the maximum speed of a small car, it would take around 114 years to reach the Sun.

The effects of gravity do not stop once you leave the Earth's atmosphere. If a rocket was fired straight upwards, once the engines stopped, it would begin to slow down, and, unless it reached escape velocity which is how fast something must travel to leave the gravitational influence of the Earth, it would start to fall straight back down again, just like a ball thrown into the air. This principle is used in sounding rockets and is described in Chapter 2 – "Rockets and Spacecraft".

To remain in space and not fall back to the Earth, a spacecraft must either go into an orbit around the Earth or be propelled so fast that it reaches or exceeds the escape velocity and the thrust away from the Earth is larger than the pull of gravity. Objects in orbit are continuously being pulled towards the centre of the Earth by gravity, in the same way that objects on the surface of the Earth are pulled downwards. In our everyday lives, we perceive the ground as relatively flat, excluding hills and mountains. However, the Earth is actually a sphere and the ground curves slightly. This can be easily proven with a clear horizon such as seen when looking out to sea. If both ends of a straight edge, such as a 30 centimetre rule, are held up in line with the horizon, the Earth can be seen to curve over the top of the rule. The Earth actually drops about five metres in every 8,000 metres or eight kilometres. If a ball were dropped from about five metres high, maybe out of a window on the third floor of a building, it would take about a second for it to hit the ground. If instead of being dropped, it was thrown horizontally, it would still take about a second to hit the ground but it would have travelled some distance along, say 50 metres. Now, if it could be thrown so that in one second it travelled eight kilometres along, it would have again dropped the five metres, but the Earth would also have curved down five metres, and so the ball would still be about five metres above the surface. If air resistance did not slow the ball down, and it did not hit anything, it would continue travelling at eight kilometres per second, dropping five metres every second, and would always remain five metres above the surface of the Earth. This is essentially how all satellites remain in orbit and do not fall back to Earth.

As the force of gravity depends on the distance between the two objects, such as the Earth and a satellite, a satellite further away from the Earth must travel slower than one closer in, to remain in a stable orbit. How fast a satellite must travel to remain in a stable orbit is called the orbital velocity, and depends on the altitude above the body the satellite is orbiting. Chapter 4 "Movement in Three Dimensions", explains more about orbits and orbital velocity. The circular orbital velocity is a balance. If it slows, the gravity of the Earth will pull the satellite downwards and it will move into an elliptical orbit, with the highest point being on the path of the circular orbit. If the lowest point of the orbit touches the atmosphere it will eventually crash into the Earth. If the satellite travelled faster than the orbital velocity the satellite would travel outwards, either into an elliptical orbit whose lowest point would be on the path of the circular orbit, or if it travelled faster than the escape velocity, it would continue off into space. The International Space Station, at an altitude of about 390 kilometres, travels at about 7.7 kilometres per second, whereas the Moon, at an altitude of about 385,000 kilometres travels at just over one kilometre per second.

It is the Earth's gravity that makes spacecraft and the Moon circle the Earth, and the Sun's gravity that makes the Earth and the other solar system planets circle the Sun. Once in a stable orbit a satellite does not need any more fuel and will keep on its trajectory indefinitely, unless disturbed by other influences, such as the gravity from other planets or drag from the atmosphere. An object that is no longer being propelled is said to coast, and the path it takes is then only influenced by gravity and its initial speed. This path is called a ballistic trajectory.

Most spacecraft coast, as, with current technology, it is only possible to propel a spacecraft continuously for a relatively short time, just as a car can only travel so far before it runs out of fuel. All of a rocket's fuel must be carried with it, and when it is coasting, it is not using fuel. If however, the spacecraft has reached the escape velocity of the Earth, it will leave the influence of the Earth's gravitational pull, and instead be pulled towards the Sun. For a satellite in orbit around the Earth to leave the influence of the Sun's gravity, it must reach about 42 kilometres per second. It would then be able to leave the solar system, and travel into interstellar space. A satellite in Earth orbit at a height of 390 kilometres, such as the International Space

Station, would already be travelling at about 30 kilometres per second relative to the Sun, as this is the Earth's orbital velocity. However, it could be up to 7.7 kilometres per second more or less than the 30 kilometres per second, depending on where it is on its orbit. If the satellite is circling the Earth from west to east, as most satellites do, it would be travelling up to 7.7 kilometres per second faster than 30 kilometres per second when it is over the Earth's night side, as it is travelling in the same direction as the Earth moves around the Sun, and slower when over the day lit side, when it is travelling in the opposite direction to the Earth's orbit. The probes *Voyager 1* and *2* and *Pioneer 10* and *11* have all overcome the influence from the Sun's gravity and are heading off into unexplored space. The *New Horizons* probe, launched in January 2006, will also leave our solar system. It is expected to pass Pluto and its moon Charon in 2015 and then continue outwards from the Sun and leave the solar system in around 2020.

Orbits

The size and shape of a satellite's orbit is mainly determined by its speed and mass and by the mass of the object it is orbiting. However, for an Earth orbiting satellite, the other planets in the solar system and the Sun and the Moon all disturb the satellite's path and so the orbit continually changes. These changes are called perturbations. The size, shape and orientation of an orbit can be described by six orbital elements or orbital parameters, which are explained in detail in Appendix A – "Orbital Elements." Two elements, the inclination and the eccentricity are used quite often throughout this book and so are also described here. The elements are defined with reference to an Earth orbiting satellite, however, they can easily be adapted for satellites orbiting other celestial bodies.

Inclination

A satellite will always travel around the centre of the body it is orbiting, and therefore it will always cross the equator twice in every orbit, unless, of course, it orbits directly above the equator. It will cut it as it travels from the southern hemisphere to the northern and

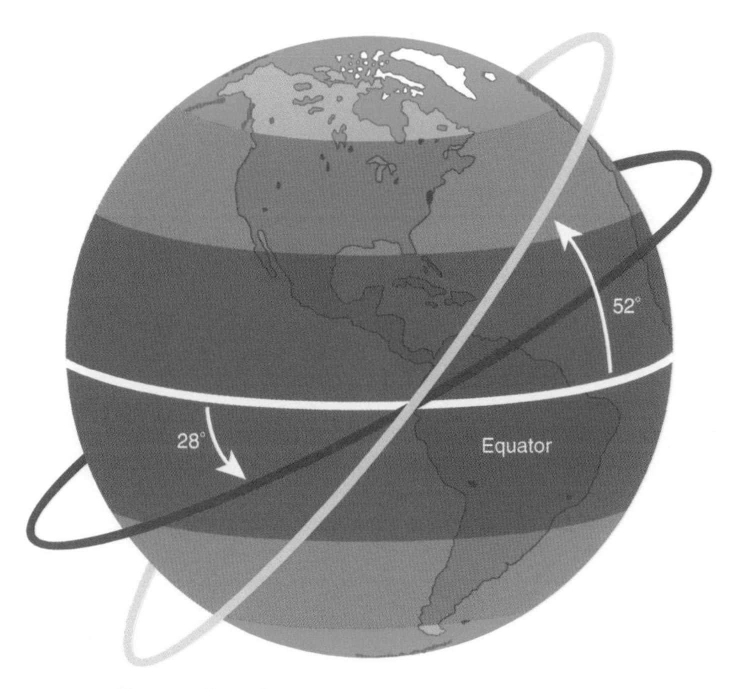

FIGURE 1.2 Satellite Inclinations.

again when it crosses from the northern to the southern hemisphere. The angle it makes when it crosses the equator from the southern hemisphere to the northern hemisphere is called the inclination, and defines the orientation of the orbit with respect to the equator. Therefore, an orbit that is directly above the equator has an inclination of 0° and one that goes directly over the north and south poles has an inclination of 90°. In Figure 1.2, the orbit shown in red has an inclination of 28°, which is the inclination of most of the USA's scientific satellites. The yellow orbit shows the path of the International Space Station, which has an inclination of about 52°.

Eccentricity

This is the shape of the orbit. Most orbits are not circular but look liked squashed circles, called ellipses. How flat the ellipse looks is called its eccentricity and it is given a value from zero to one.

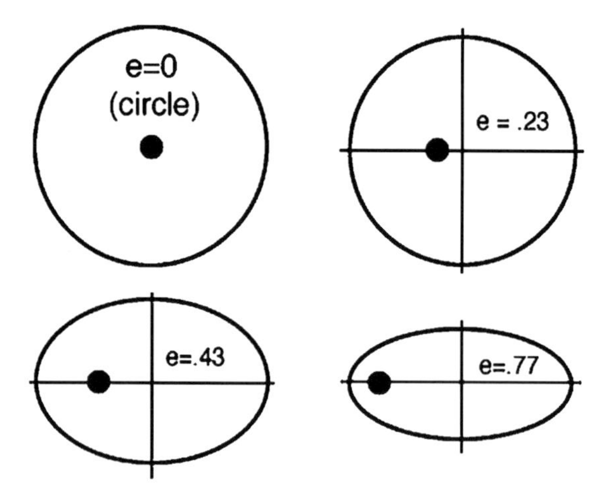

FIGURE 1.3 Ellipse Eccentricity.

An ellipse with zero eccentricity is a perfect circle and a very flat ellipse has an eccentricity nearing one, as can be seen in Figure 1.3.

Space Debris

There are a lot of bits of rock and dust in space, most of which originated from comets. As a comet passes close to the Sun, some of its surface, which is usually dirty ice, starts to melt and debris is shed in the comet's trail. Each individual dirt particle is called a meteoroid. Other meteoroids are thought to originate from the asteroid belt. Meteoroids can range in size from micrometres to about a metre in diameter. The only difference between a meteoroid and an asteroid is the size. If a meteoroid passes through the Earth's atmosphere it is heated by friction and produces light. This effect is called a meteor and is the cause of shooting stars. Usually the meteoroid burns up above the surface of the Earth. However, if a solid object lands on the Earth, it is known as a meteorite. If the Earth passes through a comet's debris field, the path of dust and ice left from the comet's tail, a shower of meteors is seen, such as the Perseids from comet Swift-Tuttle which are seen every August and the Leonids, which come from the comet Tempel-Tuttle, seen in November.

Meteoroids are not the only type of debris in space. Since the launch of *Sputnik 1* in 1957, litter has begun accumulating in the space environment. This includes rocket bodies, mission related debris, fragmentation debris and non-functional spacecraft. This space junk is called orbital debris, and is classified as small, medium and large. All objects with a diameter of less than one millimetre, or smaller than a grain of sand, are in the small category. There are a huge number of objects in this category. As they are so small, it is not possible to detect them from the surface of the Earth. Objects with a diameter between one millimetre and ten centimetres, or up to about the size of a melon, are classified as medium debris. There are estimated to be tens of millions of this type of object. Large objects are those with a diameter greater than ten centimetres and include about 700 working spacecraft and satellites. Installations across the world track about 10,000 spacefaring objects, all of which are larger than a melon, as described in Chapter 9 – "Observing Satellites".

As space debris is an international problem, the Inter-Agency Space Debris Coordination Committee (IADC) was formed. The IADC addresses the issues of space debris and is used to exchange information on research activities and to identify debris mitigation options between nations. The United Nations Committee on the Peaceful Uses of Outer Space also assesses and discusses orbital debris. Guidelines are under development, which will forbid the intentional explosion of satellites, and state that precautions must be taken against accidental events that may produce space debris. They will also suggest that decommissioned satellites must be de-orbited and destroyed in the atmosphere or, if this is impossible, which is usually the case for geosynchronous satellites, moved into a less used orbit.

Space debris or meteoroids can damage critical components on working satellites. The ultimate danger is satellites breaking into so many pieces that the rate of collisions increases enough to produce new fragments continuously. This sort of chain reaction would create a debris belt in which no object could survive.

Magnitude

The brightness of a star as viewed from the Earth is called its apparent magnitude, or usually just magnitude. The higher the magnitude, the dimmer the star will appear. Negative magnitudes are very

Table 1.2 Magnitude of Various Objects in the Sky

Object	Apparent magnitude	Comment
The Sun	-26.7	Brightest object in the sky
The full Moon	-12.7	A crescent Moon is about magnitude -9
Iridium flares	-8.0	Reflection of sunlight off solar panel on the Iridium satellites
Venus	-4.4	Often visible in the morning or evening sky
International Space Station	-3.1	Reflection of sunlight off solar panel on the International Space Station
Jupiter	-2.7	Some of Jupiter's moons are visible through binoculars or a small telescope
Mars	-2.0	Appears a red colour
Sirius	-1.4	The brightest star in the night sky
Saturn	-0.3	Saturn's rings are visible through binoculars or a small telescope
Mercury	0	Sometimes visible in the morning or evening sky, just before sunrise or after sunset
Polaris	2.0	The Pole Star
Uranus	5.5	Sometimes visible with the naked eye
Vesta	6.0	Just visible to the naked eye
Limit of human eye	6.0	Varies with different people
Neptune	7.8	Visible through binoculars or a small telescope
Pluto	15	Visible through a large amateur telescope
Eris	19	Only just visible with a large amateur telescope

bright. For example, the full Moon is about magnitude −12.7. The magnitude scale is logarithmic. Each time the magnitude reduces by one, the brightness actually increases by about 2.5.

Table 1.2 shows the magnitude of some objects visible in the night sky.

Space Law

When the first satellite was launched there were no rules governing what happened in space. When the USA and the Soviet Union set their sights on landing on the Moon, no one knew who would get there first.

Therefore they both agreed that there should be no sovereignty rights in outer space and that it should belong to all mankind. This agreement, known as the Outer Space Treaty, was ratified by the United Nations in 1967 and most countries around the world are now signatories to it. It is officially called the *United Nations Treaties and Principles on Outer Space* and sets out the basic philosophy and legal principles for outer space. It is very difficult to enforce space law and treaties are not absolutely binding, even if every country has agreed to them. The International Court at The Hague can make decisions about what can and cannot be done, but there is no method to enforce the decision. However, every nation knows what is and is not acceptable behaviour in space and the treaties are very seldom ignored.

General Information

Rocket science has developed very slowly. The Chinese probably invented rockets about 1,400 years ago. In comparison, Sir Frank Whittle's thesis on jet propulsion was written about 80 years ago, and similarly the study of atomic energy dates from Henri Becquerel's work just over 100 years ago. Rocket science has a large scope that includes many technical and scientific subjects such as engineering, mathematics, astronomy, physics, chemistry, biology, geography, meteorology, medicine and also law. Politics is also involved, as the funds used on such expensive projects need to be justified. Each subject can be further subdivided, and the amount of in-depth knowledge required to design, build and launch even a simple Earth-orbiting satellite is probably beyond any one person's capabilities or budget. The early pioneers, such as the head Soviet rocket engineer and designer Sergei P. Korolev and the German, and later US citizen, Wernher von Braun, had teams of experts helping them.

Within this book the different aspects of rocket science are covered, from the design of rockets, spacecraft and missions, to how to move and navigate in space. How the human body can survive in space and how it reacts to a weightless environment is also covered, as is a description of how to spot satellites from the Earth and a guide to the places spacecraft can go. The final chapter of the book speculates as to what the future may bring. Hopefully, this book demystifies the science behind getting man-made objects into space. It is, after all, ONLY rocket science.

2. Rockets and Spacecraft

Pre-Spaceflight

Rockets have been used for ceremonial and warfare purposes since the ancient Chinese first created them, probably in about AD 600. The first written record of the use of rockets as powered artillery was during the siege of Kai-Fung-Fu in AD 1232 , when the Chinese used rocket fire-arrows to repel the Mongol invaders. Through the 13th century other Asian armies developed gunpowder-propelled fire-arrows, and the use of these weapons quickly spread throughout Asia, Europe and the Arab countries.

Development of rockets for weapons continued and in 1647 Nathaniel Nye's *Art of Gunnery*, which included a large section on rockets, was published in London. Kazimierz Siemienowicz, a Polish–Lithuanian, also wrote about rockets and his work *Artis Magnae Artilleriae pars prima* (*Great Art of Artillery, the First Part*) was published in 1650. This contains work on rockets, pyrotechnics, fireballs and fireworks. However, the use of rockets declined as improvements were made to conventional artillery, such as cannons, by the use of better construction techniques, materials and propellants and a better understanding of ballistics.

In the early 19th century, rockets were again in use by the British. The English inventor William Congreve had developed the Congreve rocket at the Royal Woolwich Arsenal, London. These rockets were made from an iron case filled with black powder and contained either an explosive or incendiary head. The rockets were attached to wooden guide poles and were launched on simple metal A-frames. The rockets could travel up to about three kilometres with the angle of the launch frame setting the range. However, these rockets were fairly inaccurate and would often explode prematurely and were as much a psychological weapon as a physical one. They were rarely used without other types of artillery, although a barrage of thousands of rockets could prove demoralising and devastating. British ships used the Congreve rockets to pound Fort McHenry in the US War of 1812.

It is this that inspired Francis Scott Key to write of "the rockets' red glare" words in his poem that later became The Star Spangled Banner. The British Army formed a rocket brigade in 1818 and the Congreve rockets were used until the 1850s, when the English inventor William Hale removed the need for a stick by the development of a technique of spin stabilization. The exhaust gases from the rocket were made to impinge on small vanes, causing the rocket to spin. A rifle bullet in flight and a properly thrown rugby ball also use spin stabilization, as do some satellites, as described later.

Rockets were used in battle until improvements in conventional artillery again made them a less effective weapon of war. The rockets were relegated to peacetime uses and were adopted to carry rescue lines to vessels in distress and by the whaling industry, which developed rocket-powered harpoons.

The earliest known story about travelling to the Moon was written by Lucian of Samosata, in about AD 160. His first story, called *True History* or *True Story* describes Lucian and a band of sailors who sail off through the Pillars of Hercules, now known as the Straits of Gibraltar, and are blown off course by a storm. They land on an island and replenish their supplies from the rivers flowing with wine and swimming with fish. Soon after leaving the island, they are lifted up by a huge waterspout into space and are eventually deposited on the Moon, where they meet the King. It transpires that the King wants to set up a colony on Lucifer, known to us as the planet Venus, but the King of the Sun will not let him. Lucian and his sailors then join in a full-scale war and fight with such strange creatures as horse-vultures, salad-wings, millet-throwers, garlic-men and flea-archers, who rode giant fleas. He includes creatures that come from other celestial places, such as the dog-acorns who are the contingent from Sirius and cloud-centaurs from the Milky Way. Although it does sound like an early science fiction story, Lucian wrote *True History* as a satire of the contemporary and ancient literature, which portrayed mythical and fantastic creatures and events as having been real.

There is a Chinese legend that tells of an official called Wan-Hu, who assembled a rocket-powered flying chair in about the 16th century. He attached 47 fire-arrow rockets to two large kites and a wicker chair. He thought it would propel him into space. He sat in the chair, and his 47 assistants each lit a rocket. There was a huge roar and clouds of billowing smoke. When the smoke cleared, the chair and Wan-Hu were gone. Although probably apocryphal, if this story were true, it is unlikely that Wan-Hu would have made it as the first ever astronaut as the black powder rockets would not have had enough thrust to propel him very far.

The first Moon story written using the scientific knowledge of the day was by the German mathematician and astronomer Johannes Kepler, who had worked with the Danish Astronomer Tycho Brahe. Kepler also published work on optics and planetary motion, which is described in detail in Chapter 4 – "Movement in Three Dimensions". His Moon book, *The Somnium* or *The Dream*, was published after his death in 1630. It was a scientific treatise based on a dissertation he had written as a student in 1593, defending the Copernican view of the motion of the Earth around the Sun. At the time, the dissertation was not presented as the professor in charge was opposed to Copernicanism and would not allow it. The Roman Catholic Church was also opposed to Copernicanism, and maintained the Ptolemaic view that the Earth was the centre of the Universe. To oppose this view was heretical and in 1633, because of his published views for the Copernican system, Galileo Galilei was apprehended and put under house arrest for the rest of his life. *The Somnium*, when it was eventually published, was presented as a fictional narrative, with notes describing the scientific development in the story. His method of getting to the Moon was pure fantasy and involved a form of witchcraft or demon power. However, he describes a blast-off and the need for sleep-inducing drugs due to the violent shaking of the ascent, which, he thought, would be both alarming and painful. In the translation from Latin by Edward Rosen called *Kepler's Somnium*, the story reads:

> His limbs must be arranged in such a way that his torso will not be torn away from his buttocks nor his head from his body, but the shock will be distributed among his individual limbs.

Kepler therefore highlighted that high accelerations would produce unaccustomed forces on the body. Once on the Moon, he describes in detail what would be seen, including the phases of the Earth, visible only to those on the Earth-facing side of the Moon. The population on the other half of the Moon, or the far side as it is often known, would never see the Earth. He also described how long the Sun would take to cross the sky and the constellations that would be visible from different areas. *The Somnium* is considered the first serious scientific treatise on lunar astronomy.

In 1638 another science fiction story was published posthumously, this time in the English language. Francis Godwin, the Bishop of Hereford, describes in his book *A Man in the Moon* the adventures of Domingo Ganzales. Having failed to make his fortune in the Americas, Ganzales travels back to Spain only to become shipwrecked on an island. He harnesses some large birds, in the hope that they will fly him back to civilisation. The birds pull him up and up, but instead of heading to Spain, they pull him into space. There is some science included, as once in space Ganzales becomes weightless, and although the Sun was still shining he could see the stars.

These early science fiction works influenced many scientists including the Reverend Dr. John Wilkins. Wilkins wrote a book, called *The Discovery of a World in the Moon*, which he published anonymously in about 1638. In 1640, he wrote, also anonymously, *A Discourse Concerning a New Planet*. These two books were then combined and thereafter published as a single work. In it he explains, in English, some of the ideas of Galileo, Copernicus and Kepler and makes them understandable by the lay reader. He also describes Jupiter and the Moon as "worlds" and explains that they are not shining disks. This is one of the earliest popular science books.

Wilkins also speculated about the possibility of trade with the inhabitants of the Moon, whom he called Selenites. He admitted that he could not be sure that Selenites existed, but he thought they should, even though nobody had seen them. In the Discovery and Discourse books, he developed ideas about a flying chariot that could make the journey to the Moon, based on a sailing ship, a powerful spring, a clockwork gear train and a set of wings covered in feathers. He progressed these ideas in his third book called *Mathematical Magic*, which also described wind cars, guns that have multiple shots and other ingenious devices. Although some of these sounded fantastic, his books had a tremendous influence to many of the men of the early English Royal Society in the mid-17th century. At the end of the book he states:

> Yet I do seriously and on good grounds affirm it possible to make a flying chariot in which a man may sit and give such a motion unto it as shall convey him through the air. And this

> perhaps might be made large enough to carry divers [sic] men at the same time, together with food for their viaticum and commodities for traffic. It is not the bigness of anything in this kind that can hinder its motion, if the motive faculty be answerable thereunto. We see a great ship swims as well as a small cork, and an eagle flies in the air as well as a little gnat. ... 'Tis likely enough that there may be means invented of journeying to the Moon; and how happy they shall be that are first successful in this attempt.

The works of Godwin and Wilkins influenced Cyrano de Bergerac, the French dramatist whose life Edmond Rostand used as the basis for the fictional play. In de Bergerac's science fiction book *Histoire Comique des Etats et Empires de la Lune et du Soleil* or *A Comical History of the States and Empires of the Sun and the Moon*, which was published posthumously in 1656, Godwin's character Domingo Ganzales is included. Cyrano de Bergerac is credited with several ideas for space propulsion including the rocket and the ramjet.

The "Father of science fiction" is generally regarded as the Frenchman, Jules Verne. Although he was not the first to write in this genre, he was the first to make a living from it. He is noted for his works that include travel through space, air and underwater, before any practical method of getting into space had been devised and before air travel and submarines were common. His book *De la terre à la lune (From the Earth to the Moon)* was published in 1865 and the sequel *Autour de la lune (Around the Moon)* was published in 1870. These describe a three-man crew, a launch from Florida and a splashdown point in the Pacific. About 100 years later, the *Apollo* missions also had a three-man crew, a launch from Florida and a splashdown point in the Pacific.

The Russian mathematics teacher Konstantin Tsiolkovsky produced more theoretical applications towards space travel in 1903, with his writings on *The Exploration of Cosmic Space by Means of Reaction Devices*. This was a mathematical model of the use of rockets for interplanetary flight and was probably the first academic work on rocketry. Tsiolkovsky produced many papers and documents on different aspects of space travel and rocket propulsion and is considered the Russian Father of human space flight. As all his work was written in Russian, much remained unknown by other rocket pioneers for many years.

Tsiolkovsky calculated many astronautical principles and designed rockets, but he never built any. At about the same time as he was working on his theoretical models, the American scientist Robert H. Goddard began to work seriously on rocket development, although neither knew of the other's work. Goddard was much more practical than Tsiolkovsky and by 1915 Goddard had carried out his first experiments involving solid-fuelled rockets. Both Goddard and Tsiolkovsky independently came to the conclusion that the solid fuels of the time would not be sufficient to power rockets to the height they believed would make it into space, but that liquid fuels would. Liquid-fuelled rockets are a lot more complex than solid-fuelled ones and involve many parts. Goddard launched the first liquid-fuelled rocket in 1926 and by the time he died in 1945 he had been granted many patents on various component rocket parts, including combustion chambers, nozzles, propellant feed systems and multistage launchers. Some of his patents still produce royalties for his estate. Goddard is regarded as the American Father of liquid-fuelled rockets.

By the 1930s there were rocket enthusiasts and rocket clubs in many countries including Germany, the Soviet Union and the USA. The German Society for Space Travel (Verein fuer Raumschiffahrt or VfR) was formed in 1927 with the Romanian born Hermann Oberth as one of its earliest members. In 1930 the VfR successfully tested a liquid-fuelled engine and by 1932 they were regularly flying rockets. Oberth wrote his doctoral thesis *The Rocket into Interplanetary Space* in 1922, but the University of Heidelberg rejected it and he was not given his doctorate. However, he believed in his ideas and published his thesis as *Die Rakete zu den Planetenräumen (By Rocket into Planetary Space)*, which he later expanded to become *Wege zur Raumschiffahrt (The Way to Space Travel)*. Oberth is regarded as the German Father of rocketry and his books described, amongst other things, a space station and liquid-fuelled rocket designs. Oberth influenced many scientists including the young Wernher von Braun. Von Braun joined the VfR as a teenager and assisted Oberth in his spare time.

During the First World War, rockets powered by solid propellants were used as weapons. The Treaty of Versailles, the peace treaty that officially ended the First World War, forbade solid-fuel rocket research in Germany. Liquid-fuelled rockets were not specifically forbidden and, by 1932, the German Army began to take an interest in the VfR's efforts. The German Army Rocket Research Group was formed the same year, headed by Captain, later Major General, Walter Dornberger. Von Braun and most of the other members of the society eventually joined the military and the German Army Rocket Research Group. The group's main interest was to research the possibility of using liquid propellant rockets for military purposes.

With the financial support and strict requirements of the army, the scientific research and development work on rockets progressed rapidly. Von Braun, who had been fascinated with the idea of space travel and earned his doctorate in physics by the age of 22, was the technical director. By 1934, a liquid propellant rocket, named the A2, had been launched and reached a maximum altitude of 2.2 kilometres. Due to the limited availability of materials and manpower, financial constraints and rivalry between the German services, the development of the next rockets, the A3 and the A4, progressed more slowly. In 1942, the A4 was successfully launched for the first time. During one of its test flights, it reached an altitude of 189 kilometres, and was the first man-made object to be launched into space.

In his book, *V2*, Major General Walter Dornberger recalled that at the time he told his colleagues:

> We have invaded space with our rocket and for the first time – mark this well – have used space as a bridge between two points on the Earth; we have proved rocket propulsion practicable for space travel.

He continued:

> This third day of October 1942, is the first of a new era of transportation, that of space travel.

The A4 was renamed the *V2* or *Vergeltungswaffe 2* (Reprisal weapon 2). It was the first successful long-range ballistic missile and had a range of about 300 kilometres and could carry a payload of about a tonne. The majority of the design of the engine is credited to Walter Thiel and the rocket itself to von Braun. Dornberger says in his book:

> Any ambition to penetrate into space with liquid propellant rockets could be no more than wishful thinking until general technological progress provided the means for realisation. Essential prerequisites were the smelting of light alloys on a large scale, the ability to produce, and store, liquid oxygen in quantity, or

> alternatively to obtain big supplies of chemicals containing oxygen, and finally the development of electrical precision instruments.

He added:

> I think it is probable that any genuine inventor, research worker, or engineer who had had the problem to deal with under identical conditions and had worked painstakingly on scientific lines would have achieved practically the same results.

He continued:

> The time was ripe and the basic conditions were there.

He also said:

> As so often before in the history of technology, necessity in Germany after the First World War had forced a great invention to proceed by way of weapon development. Never would any private or public body have devoted hundreds of millions of Marks to the development of long range rockets purely for scientific purposes.

The first hostile *V2* fell at 6:43 p.m. on September 8, 1944, at Chiswick, near London, England. They continued to fall, mainly on London and Antwerp in Belgium, until March 27, 1945. It is estimated that the *V2* bombs killed 10,000 civilians. The *V2*s were mass-produced using mainly labour camp inmates under atrocious conditions. Over 25,000 workers died either directly or indirectly from the conditions and work of producing the bombs. The manufacture of the rocket produced more deaths than its deployment.

After the war many of the scientists and engineers who had helped develop the *V2* continued their rocket work for either the Soviet Union or the USA. Their expertise and the information gathered from unused *V2*s and other rocket parts contributed greatly to the development of the rockets that eventually launched satellites and Man into space.

Rocket Basics

Multistaging

It would be very useful if a rocket could take off from the Earth, go into orbit, come back to Earth, be refuelled and be ready to launch into space again quickly, in a similar way to an aeroplane or a car continually travelling from one place to another. However, there are technical and financial restraints that mean that, although this is theoretically possible, we do not yet have the materials or technology available to develop this type of rocket. The Ansari X-Prize, which is described in more detail in Chapter 11 – "The Future", was awarded to the first non-governmental organization to launch a reusable manned spacecraft into space twice within two weeks. As this was only required to enter space and not enter into an orbit, the winning design was only a fraction of the way to a fully reusable orbital launch system.

A technique that was utilized in the 16th century by a German firework manufacturer called Johann Schmidlap has been adopted for all current orbital space launches, although the rockets are not reusable. So that his fireworks could reach higher altitudes, Schmidlap attached smaller rockets to the top of the larger ones. When the large rocket ran out of fuel and began to fall back to the ground, the smaller one became detached, and, using its own fuel, climbed even higher. Schmidlap called this a step rocket. Today this type of system is still used, but it is known as multistaging and was independently described by Kazimierz Siemienowicz, Konstantin Tsiolkovsky, Robert Goddard and Hermann Oberth.

As the fuel is burnt, the propellant is expelled and the rocket is accelerated. The lighter the rocket, the less propellant is needed to accelerate it to the required speed to get into orbit. Or, conversely, the more propellant the rocket has onboard, the faster and further it can go. The weight of the rocket, including the engines, fuel and payload, is too large for current propulsion systems to get into orbit in one stage. Rockets therefore usually consist of separate stages. Each stage contains its own propellant, engines, instrumentation and airframe, so that it can function independently. By discarding the first stage, with its associated engine and fuel tank, the weight of the rocket is lighter and therefore the remaining stages can be more easily accelerated to the required speed.

In most rockets, the stages are stacked one on top of the other, called serial staging. The *Soyuz* launch vehicles use three serial stages, as did the *Saturn V* rockets that launched the *Apollo* missions. The stage at the bottom is called the first stage and is ignited first. The payload is usually in a protective nose cone at the very top of the stack. The first stage is the largest stage and requires the most thrust, as it must lift all of the other stages and the payload, as well as itself, off the surface of the Earth. It must also counteract the drag caused by the atmosphere. Usually, the first stage burns only for a couple of minutes. After it has used all of its propellant, the empty propellant tank, engine, instrumentation and airframe are just dead weight and are jettisoned and usually return to Earth. The second stage then ignites and further accelerates the rocket, which now has less mass.

An alternative method of staging is parallel staging, where several solid propellant motor boosters are strapped onto the side of the rocket. They form a supplementary first stage and are usually attached to the first stage. At launch, all of the rockets are ignited. The smaller rockets are sometime called the zero stages or boosters. When the strap-on rockets have used all of their propellant, which is usually before the main or sustainer engine has, they are discarded and the sustainer engine continues to burn until it too runs out of propellant. The Space Shuttle uses parallel staging. The *Titan III* and *Delta II* rockets use a combination of both serial and parallel staging. The Space Shuttle's booster rockets are salvaged after they land in the ocean and are reused, but for most other rockets, and for the Space Shuttle's main sustainer engine, the fuel tank for the first stage usually crashes into the ocean and is not recovered. Most rockets' later stages either burn up in the Earth's atmosphere or become pieces of space debris and orbit the Earth until their orbit decays and they too eventually burn up in the Earth's atmosphere.

Rockets have been designed and launched with up to five stages. There are an optimum number of stages for any rocket before adding more actually slows the rocket down. This is because the added mass and complexity of each subsequent stage counteracts the benefit of staging. As the complexity increases, the reliability decreases. It is common for rockets to use two or three stages. Each stage can incorporate many rocket engines, for different purposes. The more stages a rocket has and the heavier it is, the more expensive it is to launch. Smaller vehicles are therefore used for small payloads and low orbits. Larger ones usually have more stages, are heavier and are therefore more expensive, but can carry larger payloads or take them to higher orbits or even out of Earth orbit altogether.

Launch Pad to Orbit

The steps to get a rocket from the launch pad and into orbit are: pre-launch, lift-off, vertical rise, pitchover, ascent, staging (where the first stage runs out of propellant, called burnout, and is jettisoned and the second-stage ignites), ascent, staging (where the second stage reaches burnout and is jettisoned and the third-stage ignites), ascent and final burnout.

The pre-launch phase begins when the rocket arrives at the launch site. This is usually a couple of weeks before the launch. This phase ends when the countdown starts. The first stage is used to lift the rocket off the Earth's surface and carry it vertically upwards to clear the launch tower. This usually takes a few seconds and is known as the vertical rise time or VRT. After the launch tower is cleared the exhaust nozzles rotate slightly, so the ascent is slightly off-vertical. This is called a pitchover manoeuvre. To reach space the rocket must overcome both gravity and drag caused by the atmosphere. The drag can be minimized by flying straight upward and getting out of the atmosphere as fast as possible. However, to get into an orbit around the Earth, the rocket must travel horizontally. If it carried on upwards, it would eventually run out of fuel and the Earth's gravity would pull it back down, just like a stone thrown straight up. Therefore the rocket pitches slightly and the angle of the flight path increases steadily towards the horizontal as the Earth's gravity pulls on it. This is called a gravity turn. If this first part of the flight is disturbed, maybe by the force of the wind or by the propulsion system being misaligned, guidance and control systems determine the amount of variation from the planned flight path and make any necessary corrections, so that the vehicle gets back on course.

The rocket keeps climbing and accelerating as the different stages reach burnout, and the next stage takes over. The protective nose cone is usually jettisoned during the second stage burn as the payload no longer requires protection from the wind after it is out of the lower atmosphere. When the required altitude, orbital speed and direction are reached, the second stage is either shut down or it reaches burnout. If the payload is destined for an orbit higher than about 1,000 kilometres or is leaving the Earth's orbit, a three-stage rocket will usually be used. The second stage will have placed the third stage and the payload into a low Earth parking orbit. This is a temporary orbit, where the spacecraft can wait for the correct timing before the third stage fires and moves it into its final orbit or another trajectory. This waiting time, or coast period as it is known, is usually between 30 minutes and an hour, but it can be longer depending on the mission. The propulsion system for this final manoeuvre may be integrated with the payload or it may be discarded when it is used.

Once the launch vehicle has released its payload, it has no further useful purpose and remains circling the globe in tighter and tighter orbits until it eventually burns up in the Earth's atmosphere. In September 2006 there were over 6,500 spent rocket bodies and other pieces of debris orbiting the Earth.

Once in orbit, the type of propulsion system can be changed. This is because leaving the Earth requires the rocket to be accelerated quickly through the atmosphere and around the planet before it falls back to it. Once in orbit, the pull of gravity from the Earth is balanced by the speed of the spacecraft around the planet, and the spacecraft does not fall back to Earth and so slower accelerations can be used to change the path of the trajectory. This allows the use of much more efficient motors that produce more thrust per quantity of propellant, such as ion drives. These are explained in Chapter 5 – "Propulsion Systems".

Launch Vehicles

The launch vehicle is the rocket, including all of the stages, that is used to launch a payload into space. The structure consists of the fuel or propellant tanks, a frame onto which the propulsion systems are mounted and an aerodynamic shroud, which provides a low-drag housing for the rocket and all of the components. It also contains all of the guidance and control systems that are needed to put the payload into the required orbit and the payload canister, where the payload is stored during the launch until it has reached orbit. The part of the shroud protecting the payload canister is called the payload shroud, the payload fairing or the nose cone. The whole rocket has tight weight limitations and therefore it is made with the least amount of material that will withstand the severe stresses or loads encountered both on the ground and during the flight. The ground loads include supporting its own weight while on the launch pad. It must also be able to withstand any force from the wind, which is sometimes helped by the use of spoilers or guy ropes. If the wind is predicted to get too strong when the Space Shuttle is on the launch pad, the Space Shuttle is removed and rolled back to the protection of the Vehicle Assembly Building. The flight loads include the force due to the acceleration, drag as it passes through the atmosphere and forces caused when the direction of travel is changed. This includes any liquid propellant sloshing about within the fuel tanks.

The preliminary design decisions for building a launch vehicle therefore include how many stages are to be used, the type of propellant for each stage, any size limitations and what materials will be used. The vehicle should be able to achieve the required orbit or trajectory with the maximum reliability but at the minimum cost. The mass of the launch vehicle, fully fuelled and ready to go, divided by the mass of the vehicle when all the propellant has been burnt, is called the mass ratio. A major consideration in rocket design is to keep this mass ratio as high as possible. The lighter the launch vehicle is, the more it can take up into space, or the further it can travel when it is up there. Other problems that must be considered in the design of the airframe include vibration and fatigue. The structure will flex and bend slightly from the various forces, which, in themselves, may not cause failure, but over time may cause problems, just as a metal paperclip repeatedly bent and unbent will eventually break in two. The environmental conditions that the structure must withstand are also important considerations. The vacuum of space can cause materials to evaporate and the extremes of temperature may cause them to melt or become brittle. The possibility of meteoroid impact is also of concern, as this can damage critical components or even cause an explosion. The amount of radiation the spacecraft will pass through must also be considered as it can damage electrical components, sensors and also items in the payload, particularly living items such as plants, animals or humans.

The thrust or impulse produced by an engine changes the speed or velocity of the rocket. The amount of thrust a launch vehicle can produce determines how high the rocket can get and how much it can lift, and therefore determines the mass and orbital height of the payload. Currently, the only way to produce enough power to launch a spacecraft from Earth is by the combustion of chemical propellants. The different types of propellant and propulsion methods are described in detail in Chapter 5 – "Propulsion Systems".

Once a launch vehicle has been proven to be reliable, it is usually modified so that it can take a wide variety of payloads. For example, the European Space Agency's (ESA's) *Ariane 2* and *Ariane 3* were almost identical, but *Ariane 3* was equipped with strap-on boosters and so it could take a heavier payload or take it to a higher orbit, as can be seen in Figure 2.1.

FIGURE 2.1 Launch Vehicles *Ariane* 2 and *Ariane* 3.
Image courtesy ESA

The choice of launch vehicle for each different payload is determined by numerous factors. These include the environmental conditions that the payload can endure, such as vibration, electromagnetic radiation and amount of acceleration. Launch vehicles that propel a manned crew are limited in the rate with which they can accelerate, as a human crew can only survive a certain amount of force, as described in more detail in Chapter 8 – "Humans in Space". The size of the payload is limited by the size of the payload canister and so items such as antennas and solar arrays are usually compacted for launch and take on the desired shape only after they have left the protective nose cone. Once released, the payload is generally known as a spacecraft and consists of the instrumentation for the mission and the spacecraft bus. The spacecraft bus looks after all the housekeeping for the spacecraft, such as attitude control, power supply and communication.

Sounding Rockets

A sounding rocket carries instruments into the upper atmosphere or into space. This type of rocket has a sub-orbital flight path, meaning that it does not make a complete orbit of the Earth. The rocket is launched almost vertically and once it has used all of its fuel, it releases its payload. The payload descends back to Earth in an arc-shaped path. Sounding rockets have been used to take instruments over 1,500 kilometres above sea level, although they are more commonly used between the ranges of 50–200 kilometres. This is the region that is too high for balloons and too low for satellites to gather data. The instruments are designed to take measurements and perform scientific experiments during the flight, and more than one payload can be launched from a single launch vehicle. Parachutes are usually deployed to slow the fall so that the instruments and data can be recovered. The instruments can then be reused, saving scientists both time and money. As the launch rocket does not need to put the payload into an orbital trajectory the amount of thrust required, and therefore the corresponding cost, is lower than for launching a satellite. The payloads can still be quite large though as payloads of over 500 kilogram have been flown. The origin of the term "sounding rocket" comes from the nautical term to sound. This was the process of measuring the depth of water, but the term now covers the determination of any physical property at a depth in the sea or at a height above the Earth.

The Ansari X-Prize challenge was effectively a manned sounding rocket. Details about the winner of this prize are given in Chapter 5 – "Propulsion Systems". Although the challenge was difficult, it requires a lot less energy for a sub-orbital flight than for attaining an Earth orbit. On May 5, 1961, the first American manned spaceflight, with Alan B. Shepard aboard the *Freedom* 7 spacecraft, was a sub-orbital flight.

Attitude Control and Movement

The attitude, or direction the spacecraft is pointing, must be controlled and kept stable once in space. This is so that the communications antenna can be accurately pointed towards the Earth and any onboard instruments can obtain data from known directions. Also, propulsive manoeuvres used to change the orbit must be conducted when the satellite is orientated in the correct direction, otherwise the new orbit will be incorrect. Some instruments or equipment, such as solar panels and optical equipment, need to be able to move independently in relation to the main body of the satellite and so accurate attitude determination is essential. The method of calculating the attitude is explained in detail in Chapter 6 – "Navigation in Three Dimensions".

While a rocket is still within the Earth's atmosphere, fins can be used to control the direction of flight. However, once outside the atmosphere, fins will no longer work and to influence the flight path or the attitude of the spacecraft a change in the direction of the thrust is required. This change is achieved by angling the exhaust nozzles. Once in freefall, the rocket's attitude has no influence on the path.

Gravity Gradients and Tidal Forces

A gravity gradient is caused by the change in the strength of gravity with the distance between two bodies, such as the Earth and the Moon. If all other forces are equal, it will cause any orbiting object, whose mass is not evenly distributed, to twist around so that the more massive end points towards the body it is orbiting. For example, if a bowling pin were to orbit the Earth, after time, the base would point towards the Earth, and the top would be pointing away from the Earth. As the Moon is in orbit around the Earth, the parts of the Moon closest to the Earth receive a slightly stronger gravitational attraction than those parts further away. Although the Moon looks round and symmetrical, the mass within it is actually unevenly spread and the side we see from the Earth contains more mass than the side we cannot see. It is the gravity gradient that keeps one face of the Moon always pointed towards the Earth, and the far side always pointed away. This is also known as tidal locking. A gravity gradient also causes the tides. The Moon does not orbit the centre of the Earth, but rather the centre of mass of the Earth–Moon system, called the barycentre. This is located about 1,700 kilometres below the surface of the Earth, on the side facing the Moon. As the Earth rotates daily, the barycentre is not at a fixed location in the planet. The Earth also revolves around the barycentre as it orbits the Sun. At the centre of the Earth, the centrifugal force produced by the orbits of the Moon and the Earth around the barycentre balances the gravitational forces pulling the Earth and the Moon together, so the two bodies remain the same distance apart. However, at the surface of the Earth the two forces are not equal. On the side of the Earth nearest to the Moon the gravitational pull from the Moon is stronger and the oceans are pulled towards the Moon, which forms a bulge. On the side of the Earth opposite to the Moon the gravity is weaker and the centrifugal force pushes the water in the oceans out into another bulge in the opposite direction to the Moon. These two bulges are the cause of the tides, although other factors such as the daily rotation of the Earth and the position of the landmasses complicate the effect. The crust of the Earth and the atmosphere also move in response to the gravity of the Moon. The surface of the Earth can rise and fall up to about 20 centimetres with each tide and this can trigger earthquakes and volcanoes. The Sun also has a tidal effect on the solid Earth, the atmosphere and the oceans. This leads to the variation in the tides, known as springs and neaps, when the Sun and the Moon are either aligned or not aligned, in relation to the Earth. The Sun is 27 million times more massive than the Moon,

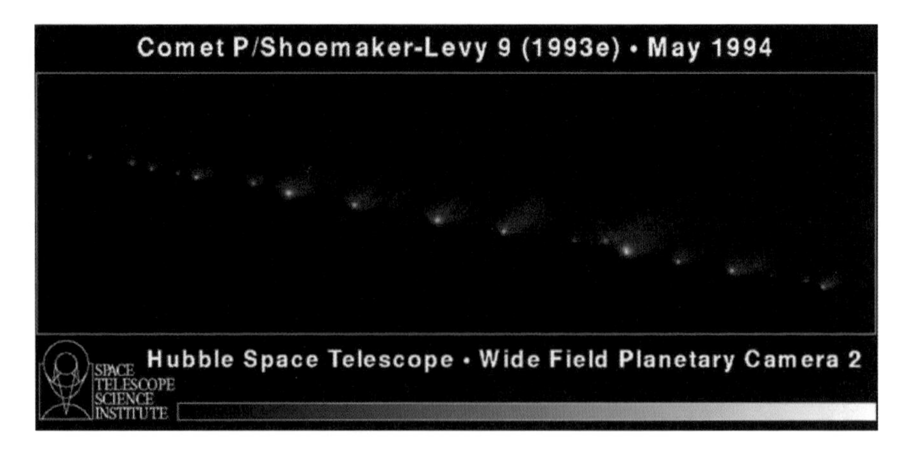

FIGURE 2.2 Break-up of Comet Shoemaker Levy 9.
Image courtesy H.A. Weaver, T.E. Smith (Space Telescope Science Institute), and NASA

but is about 390 times further away. Although gravity is inversely proportional to the square of the distance between two objects, the inverse square law, as described in Chapter 1 – "Introduction", the tide generating force is not and instead it is inversely proportional to the cube of the distance between the two objects. This means that although the effect of the Sun's gravity on the Earth is larger than the Moon's, and therefore we orbit the Sun, the tidal effect from the Sun is a little less than half that of the Moon. Jupiter's gravity gradient caused the breakup of the comet Shoemaker-Levy 9 before it hit the planet in May 1994, which can be seen in Figure 2.2.

Once a satellite is in orbit external forces such as the gravity gradient can cause changes in its momentum and therefore cause changes to its attitude and orbit, called perturbations. Solar radiation pressure can also cause perturbations. Sunlight is made up of tiny packets of light called photons. These photons carry momentum when they move, which can be transferred to other objects. This is called solar radiation pressure. Applying an impulse or thrust to the spacecraft counteracts the influence of these external forces. This manoeuvre is called a momentum desaturation (desat) or momentum unload manoeuvre and is usually performed by thrusters expelling propellant in the desired direction.

The gravity gradient can also be used to stabilize satellites. Long booms are sometimes deployed from the satellite after launch and the difference in gravity from the point nearest the Earth to that furthest away keeps the satellite aligned. This system can only be used in low Earth orbit, as any higher and the gravity is too weak. It was first used on *Transit*, the Navy Navigation Satellite System, in the early 1960s.

Spin Stabilization

If an object, such as a spinning top or bicycle wheel spins at a sufficient rate, it will resist any perturbing forces and remain pointed in more or less the same direction. This principle was used to stabilize some of the early satellites such as NASA's *Pioneer 10* and *11* spacecraft and also *Telstar*, the first active communications satellite. It is still used to stabilize satellites today, including some weather satellites and the relay satellite for the Japanese Space Agency's *SELENE* lunar orbiter mission. Spin stabilized satellites are generally cylindrical, like a tin can, and usually spin once every second, although larger and heavier satellites require a faster spin rate. The spin rate can be altered by the use of propulsion thrusters. A major disadvantage of this type of stabilization is that large flat solar arrays cannot be used to obtain power from the Sun, instead, solar cells are wrapped around the cylinder, like a label around a tin of soup. Large batteries are used to store the power. ESA's meteorological satellite, *Meteosat Second Generation (MSG)* uses spin stabilization and wraparound solar cells, as can be seen in Figure 2.3. Another problem is encountered if instruments or antennas are required to point at specific places. In this case the satellite can be split into two linked bodies, one is spun to keep the satellite stable and the other is despun so the instruments and antenna can remain pointed at their targets. Spin stabilized spacecraft are most suitable for measurements such as magnetic fields and charged particles, as they provide a continuous sweeping motion.

Three Axis Stabilization

By being able to change the direction of each of the three axes on a satellite, any attitude can be obtained. The three axis stabilization

FIGURE 2.3 Artist's Impression of ESA's Meteosat Second Generation (MSG) Spin Stabilized Satellite.
Image courtesy ESA

is achieved by nudging the spacecraft back and forth using small thrusters, called a reaction control system or RCS, or by using reaction wheels. A reaction control system is also known as an auxiliary rocket propulsion system. This usually consists of 12 nozzles that can provide low thrust in each of the six degrees of freedom of up–down, left–right, forwards–backwards, roll, pitch and yaw, as described in Chapter 4 – "Movement in Three Dimensions". Liquid propellants, cold gas systems and electrical propulsion systems have all been used in RCSs. Large thrust RCSs can also be used during the launch and in space to correct the trajectory of the spacecraft. The disadvantage of the RCS is that when the propellant has run out the satellite can no longer be controlled and it becomes useless. *Voyager 1* and *2* have used a RCS for attitude control since 1977. By April 2006, they had used up a little over half of their 100 kilogram supply of propellant. Some satellites contain electrically powered spinning wheels, called reaction wheels or momentum wheels. These are very heavy wheels, mounted at right angles to each other. If the sensors on the satellite detect that the satellite is not in the correct orientation one, or a combination, of the spinning wheels either speed up or slow down. Slowing down a reaction wheel will turn the spacecraft in one direction, speeding it up will turn it in the opposite direction. Reaction wheels help to provide a steady spacecraft for observations and measurements but they do add mass to the spacecraft, which is expensive to launch. They also have a limited lifetime. The major advantage of three axis stabilization is that instruments and antennas can point in any desired direction without having to be despun.

Magnetic Torquers

Magnets interact with the Earth's magnetic field. If they are attached to a long boom connected to a satellite they will cause a torque or twisting force on the satellite. This torque can be used to align the satellite in a similar way to a compass needle always pointing north. This type of attitude control is called a magnetic torquer. By using three electromagnets at right angles to each other and by changing the electrical current supplied to them the amount of torque and therefore the attitude of the satellite can be controlled. However, if a satellite is in orbit directly over the equator, the Earth's magnetic field would always lie south to north and the magnets cannot easily be used to realign the satellite in the north–south direction. If the satellite is in a polar orbit, all torque directions can be achieved, as at some point in the orbit, the Earth's field direction in relation to the satellite will be in a suitable direction. This type of stabilization only requires electricity and it therefore has no exhaust, but it can cause magnetic disturbance to electronics within the craft. It can be used up to geosynchronous altitudes although the strength of the magnetic field weakens with the altitude above the Earth. The strength and direction of the field also changes slightly across the globe and so a magnetometer is usually used to measure the local conditions. As the *Hubble Space Telescope* has sensitive optics that any exhaust could contaminate, it uses magnetic torquers during its desat manoeuvres.

Spacecraft

There are almost as many different types of spacecraft as there are missions, and each spacecraft is designed to fit certain criteria associated with a particular launch vehicle. These criteria include mass, size and also the tolerance to forces such as acceleration and vibration. For example, the *Hubble Space Telescope* was so large the only way it could be taken safely into space was in the cargo hold of the Space Shuttle. Each spacecraft can usually fit into one of the following broad categories:

Communication and Navigation Spacecraft

There are many spacecraft of this type in Earth orbit, including the Global Positioning System or GPS satellites, the first German TV broadcasting satellite, *TV-SAT A3*, and satellite phone communication systems such as the *Iridium* constellation. These types of satellites are often in a geostationary orbit, but, depending on their role, they are not confined to such an orbit. In the future, communication and navigation spacecraft may be deployed around other planets such as Mars or Venus. These would be dedicated to communicate with other spacecraft in their vicinity. Currently, various orbiter spacecraft are equipped for limited communications relay and do this to some extent but they are not dedicated communication spacecraft. For example, *Huygens*, the probe that went to Saturn's moon Titan, used the spacecraft *Cassini* that is in orbit around Saturn to relay data back to the Earth in 2005. Earth Observation Satellites, which include scientific spacecraft that track the environmental conditions on the Earth, such as ESA's *Envisat*, and also weather satellites fit into this group.

Fly-by Spacecraft

As their name suggests, these spacecraft fly by objects in the solar system but do not get caught in a planetary orbit, and so either continue in a solar orbit or leave the solar system on an escape trajectory. This type of spacecraft was used to give us much of our early information about the solar system, such as from the spacecraft *Mariners 2* and *5* that went to Venus, *Mariners 4, 6* and *7* which went to Mars and *Mariner 10* that went to Mercury. *Pioneers 10* and *11* which flew past Jupiter and Saturn, and *Voyager 1* and *2* which between them flew past all of the gas giants – Neptune, Uranus, Jupiter and Saturn – are all now approaching interstellar space. The *New Horizons* mission to Pluto and the Kuiper Belt is also a fly-by mission. These spacecraft must be capable of using their instruments to observe the targets as they pass them. Their optical instruments pan to compensate for the apparent motion of their target. The data they generate is stored onboard until their antennas point towards the Earth, when they can download their information. They must be designed to survive long periods of interplanetary cruise, usually with their instruments protected from damage from micrometeoroids and solar radiation.

Orbiter Spacecraft

These spacecraft are designed to travel to another planet or object and orbit around it. They are currently continuing the exploration of our solar system, with in-depth studies of some of the planets. The *Cassini* Saturn Orbiter, *Mars Reconnaissance Orbiter* and the *Ulysses* Solar Polar Orbiter are all orbiter spacecraft. This type of spacecraft must still be able to function when they are in the planet's shadow, when they will cool down substantially and will not be able to generate electricity from solar arrays. They must be able to save data when the Earth is not in direct view, and send the information later. The design must also allow the spacecraft to decelerate substantially at the right moment, so they can achieve orbital insertion.

Atmospheric Spacecraft

Atmospheric spacecraft are designed to collect data about the atmosphere of a planet or natural satellite. They are typically relatively short missions and are generally carried to their destination by another spacecraft. *Huygens*, which was carried to Saturn's moon Titan by the *Cassini* spacecraft in 2005, was an atmospheric spacecraft. This type of spacecraft usually needs an aeroshell, an aerodynamic braking heat shield, to slow them down and protect them from the heat created by atmospheric friction during atmospheric entry. The use of an aeroshell is called aerobraking. After the aeroshell is jettisoned, these spacecraft need parachutes or retrorockets, rockets that are used to slow the motion of the craft, so that they can descend slowly. The scientific instruments onboard usually take measurements of the atmosphere's composition, temperature, pressure and density. Some atmospheric spacecraft land on the surface and continue to send back data, and so can also be classified as landers.

Lander and Rover Spacecraft

Lander spacecraft are designed to reach the surface of a planet and survive long enough to send the data back to Earth. The Soviet *Venera* landers in the 1960s managed to survive the harsh conditions on Venus long enough to carry out chemical composition analyses of the rocks and relay colour images. NASA's *Surveyor* series of landers carried out similar experiments on the Earth's Moon, also in the 1960s. Rover craft move about on the surface of the planet and gather more information. They are semi-autonomous as the delay in radio communication over interplanetary distances means they must be able to make some decisions on their own. They are usually used for taking images and analysing soil and rocks. The *Mars Exploration* Rovers, *Spirit* and *Opportunity*, which landed on Mars in 2004, are probably the most well-known rovers.

Observatory Spacecraft

These spacecraft do not travel to a destination to explore it. Instead, they observe distant targets from either an Earth or a solar orbit, without the obscuring and blurring effects of the Earth's atmosphere getting in the way. Examples include the *Hubble Space Telescope*, the *Chandra X-Ray Observatory* and the *Solar and Heliospheric Observatory, SOHO*.

Penetrator Spacecraft and Impactors

Spacecraft that have been designed to penetrate the surface of a body, such as a comet or asteroid, are called penetrators. Once they have survived the landing, they then take readings of the properties of the object. This data is usually then sent to an orbiting spacecraft and relayed to the Earth. Impactor missions gather data by impacting the surface and analysing the results of the impact. In 2005, NASA's *Deep Impact* was a fly-by spacecraft that fired an impactor into the interior of the comet Tempel 1, thus excavating debris from the interior. The fly-by spacecraft *Deep Impact* and the Earth orbiting *Hubble Space Telescope, Spitzer Space Telescope* and *Chandra X-ray Observatory*, all recorded the impact. The images showed the comet to be more dusty and less icy than expected. As the impact generated a large, bright dust cloud the impact crater was obscured from view. Also in 2005, the Japanese *Hayabusa* spacecraft successfully landed a probe on asteroid Itokawa. Hopefully, it managed to take a sample of the asteroid, by firing a bullet or impactor into the asteroid and catching any debris that was thrown up. The probe then returned to its spacecraft, which is now on its return journey to the Earth. However, there were a few technical problems and the sampler may not have been successful and communication with and control of the spacecraft has become difficult.

Manned Spaceflight

The first human carrying spacecraft was *Vostok 1* on April 12, 1961. It carried the Soviet cosmonaut Yuri Gagarin once around the Earth. Since then over 200 spacecraft carrying humans have been launched. Most of these spacecraft have been either the *Soyuz* or the Space Shuttle. Spacecraft that carry a human crew and passengers have more design constraints than unmanned spacecraft. It does not matter if the craft is 15,000 metres or 15,000 kilometres above the surface of the Earth, humans need a sealed pressurized cabin containing an atmosphere that is approximately the same as normal conditions on the Earth. More details about living in space are included in Chapter 8 – "Humans in Space".

Power Systems

If the electric power on a spacecraft fails, the usefulness of the mission will usually end. *Sputnik 1*'s batteries ran out of power three weeks after it was launched. It therefore stopped transmitting its distinctive bleep and with it its encoded temperature and pressure readings. *Sputnik 1* carried on circling the Earth, until it burnt up in the Earth's atmosphere 92 days after launch. Since then, the demand for power in spacecraft and satellites has increased greatly. *Vanguard 1*, launched by the USA in 1958, required about 1 watt of power. This is a hundredth of the power required to light a 100 watt lightbulb. ESA's *Mars Express*, launched in 2003, has a maximum power requirement of about 500 watts, the equivalent to five 100 watt lightbulbs. The Space Shuttle's onboard systems uses over 100 times that of *Mars Express* as it requires 5–10 kilowatts of power, and the International Space Station uses about 100 kilowatts for onboard systems. A spacecraft's power system can be split into the primary and secondary energy source and the power control and distribution network.

Primary Energy Source

The primary energy source is the main power source for a spacecraft. Electrical power is usually generated from the conversion of sunlight or of a fuel. The type of power source used is dependent on the amount of power required and also on the length of the mission. If the power requirement is very small and for a short duration, batteries may be used to provide stored energy.

Solar Arrays

Most spacecraft use solar panels or arrays as their primary energy source. These convert energy from the Sun into electrical energy. An array is made from many thousands of individual solar cells connected together and can provide power levels from a few watts to tens of kilowatts, depending on the size of the array. The solar cells are made from materials that are semiconductors, such as silicon or gallium arsenide. By adding small amounts of chemicals such as boron and phosphorous to the semiconductors, they will become either electron deficient, in which case they are called a p-type material, or they will have an electron excess, called a n-type material. When the cell is illuminated sufficiently a voltage difference is created between the n-type and p-type materials and electrical energy is produced. This is called the photovoltaic effect. By connecting different numbers of cells together, the power supplied can be varied to suit the requirements. The larger the surface area of solar cells that is pointed towards the Sun, the more power is generated. However, getting large solar arrays into space is problematic. As the payload space is limited, solar arrays have to be packed away during the launch and deployed once in orbit. Once deployed, they are usually pivoted so that as the spacecraft moves, they are always facing the Sun, no matter the orientation of the spacecraft. The energy produced by the solar arrays is usually stored in batteries. If the batteries are full, or the spacecraft requires less electricity than the amount generated, the solar arrays are often turned so they do not point directly at the Sun and therefore they produce less electricity. Any excess power that is made and not required and cannot be stored in batteries is vented off and is usually dissipated as heat and radiated into space.

Solar power is not usually used as a primary energy source for spacecraft operating farther from the Sun than about the orbit of Mars, which is about 1.5 Astronomical Units (AU) or one and a half times the distance the Earth is from the Sun. This is because the amount of power that can be derived from the Sun diminishes with the distance away from it. However, if the mass of a solar array could be reduced, or the power generated per unit area could be increased, solar arrays operating further away from the Sun may become feasible. The furthest that solar arrays have so far been used is two Astronomical Units by NASA's *Stardust* spacecraft. ESA's *Rosetta* space probe, launched in 2004, uses solar panels and its trajectory will take it out as far as the orbit of Jupiter which is 5.25 Astronomical Units, however, it will spend most of its mission closer to the Sun. The *Magellan* mission to Venus and the *Mars Global Surveyor* used solar power, as does the *Hubble Space Telescope* and the International Space Station (ISS), which are in orbit around the Earth.

Fuel Cells

A fuel cell generates electrical energy in a similar way to a battery, and as long as it is supplied with fuel it will continue to produce electricity almost indefinitely. A battery, however, will run down and require recharging. The electricity from a fuel cell is produced by the chemical reaction of a fuel, such as alcohol, and an oxidant, such as oxygen. The fuel cell consists of two electrodes sandwiched around an electrolyte, a substance that is capable of conducting an electric current. The fuel passes over one electrode and the oxidant over the other. Electrons are stripped from the fuel and the flow of these electrons from one electrode to the other generates electricity that can be harnessed. Once the fuel is stripped of some of its electrons, it becomes an ion, which passes through the electrolyte. At the second electrode, the ions recombine with the electrons and also with the oxidant, to produce a reactant or waste product. There are many different combinations of fuel and oxidant possible. In space a hydrogen cell, which uses hydrogen as the fuel and oxygen as the oxidant, is often used. The Space Shuttle uses this type of fuel cell and the astronauts can then use the waste product, which is water. Other fuel cells use hydrocarbons and alcohols as the fuel, and air, chlorine or chlorine dioxide as the oxidant. The fuel and oxidant do not burn as they do in an internal combustion engine, and therefore fuel cells are virtually pollution free. Another advantage is that they have no moving parts that could wear out. As the fuel is converted directly to electricity without burning, a fuel cell can produce more electricity from the same amount of fuel than an internal combustion engine and it is therefore more efficient.

Nuclear

Radioisotope Thermoelectric Generators

A radioactive isotope, or radioisotope, is a material that, due to processes within its atomic structure, decays and gives off energy mainly in the form of heat. Some radioisotopes decay quicker than others and they therefore provide more power or energy per second.

Radioisotope thermoelectric generators (RTGs) produce electrical power from the heat generated by radioactive decay. The heat is converted into electricity by the use of thermocouples and heat sinks, or thermoelectric converters. Thermocouples are made from wires made from two conductors of different materials, such as copper and iron, that are connected at two joints to form a continuous loop. If one joint is kept warmer than the other, an electrical current will flow around the circuit. This is called the Seebeck effect. If a semiconductor, such as silicon, is used instead of a metal wire, the current will be increased. On a RTG, the nuclear fuel heats one joint of each thermocouple and the other joint is connected to a heat sink. The heat sink works like a radiator. It removes the generated heat from the thermocouple and keeps the two joints at different temperatures.

RTGs have been used to provide electrical power in space since 1961. *Pioneer 10*'s RTGs have been working for over 30 years and continue to provide power as it travels out beyond the orbit of Pluto and towards the edge of the heliosphere. The *Apollo* missions used RTGs, as does the more recent *Cassini* mission to Saturn. The main advantage of RTGs is that they can provide power over very long time periods. This makes deep space missions possible, which would not be feasible with the use of solar arrays as the light from the Sun would be too dim, or with fuel cells as insufficient fuel could be carried. They are also suitable for missions that are not in view of the Sun for long periods, such as lunar landers. However, they do affect the radiation environment around the satellite and can affect instruments and equipment. To decrease the radiation near the spacecraft, the RTG may be mounted on a long boom. RTGs also produce a possible radiation hazard during the building of the satellite, and could cause further problems if there is a launch failure. RTGs are usually used to power internal power systems of spacecraft, however, they can also be used to provide electricity for an electric propulsion system.

Nuclear Fission Reactors

Nuclear fission is when an atom is split into several smaller fragments, all of which are about the same size. If the mass of all the fragments were known, they would add up to less than the mass of the original atom. This "missing" mass is about 0.1% of the original mass and is converted into energy, usually as heat. This heat can be used to generate electricity. Conventional ground-based nuclear power stations use nuclear fission to generate heat. The heat is used to turn water into steam, which is then fed through turbines that turn generators that produce electricity. In space based nuclear fission systems, the principle is very similar, but with the use of a thermoelectric converter, as described above, rather than turbines. The former Soviet Union flew over thirty nuclear fission reactors, including the low powered fission reactors in their *Radar Ocean Reconnaissance Satellites (RORSATs)*. The USA has flown one fission reactor, the *SNAP-10A*, which was aboard the *SNAPSHOT* spacecraft in 1965.

Solar Heat

The heat from the Sun can also be used to produce electricity in a similar way to the RTG and solar fission reactors, in addition to the method used by solar arrays. This type of system is called solar thermoelectric. It can also be used to power a turbine which then generates electricity, and this is called solar dynamic. NASA has investigated various solar dynamic systems since the 1960s. It was originally intended that some of the International Space Station's power would be generated by a solar dynamic system, as, over time, it had cost advantages over photovoltaic systems. However, as the development costs have proved high and there has been no solar dynamic system used in space so far, a conventional photovoltaic and battery system was chosen instead.

Secondary Power Source

When the primary power source is not available, for example, if the solar arrays are in the Earth's shadow, a secondary power source is required. These are usually batteries, although some systems use a combination of solar arrays and fuel cells. *Sputnik 1* relied on non-rechargeable batteries for its only power supply, but with improvements in technology, rechargeable batteries are usually used as the secondary power source.

Batteries release energy that is stored in chemicals by an oxidation–reduction process. There are four main constraints that limit the type of battery used. These are the amount of power required, the mission lifetime, the system mass and the cost. Also, where the spacecraft is, and therefore how long it must rely on its secondary power source, determines which battery type to use. The major differences between battery types are the electrical, charge, discharge and lifetime characteristics.

The electrical characteristics are governed by the oxidation–reduction reaction that takes place within each cell. This influences the nominal voltage or average voltage that each cell produces, the capacity, the operating temperature and also the energy density. The capacity is the amount of electricity, measured in amperes, which the battery can supply in one hour for a total discharge. For example, a 20 ampere hour battery can supply 20 amperes for one hour or five amperes for four hours. The operating temperature is the range of temperatures at which the battery performance varies least and is directly related to the temperature range at which the oxidation–reduction reaction is quickest and most efficient. If the temperature drops below this range, the voltage, current and energy density will drop off. Should the temperature rise above this temperature range, the chemicals within the cell may decompose or even explode. The battery's energy density represents the amount of energy, measured in watt hours, that can be stored in the system, per unit mass or volume.

The charge and discharge characteristics of the batteries are important on missions with quick discharge and recharge times. For example, a satellite in a low Earth orbit may experience an eclipse nearly 16 times a day. The total eclipsed time is about equivalent to the time spent recharging. The battery therefore needs to be able to be discharge and recharge relatively quickly. The amount the battery discharges between recharges is called the depth of discharge or DOD. Large DODs cause the lifetime of the battery to be short. The battery lifetime is the number of discharge–recharge cycles that the cell can withstand before it fails. The requirements of a satellite in a low Earth orbit mean that nickel cadmium (NiCd) and silver zinc (AgZn) batteries are often used, whereas in a geosynchronous orbit nickel hydrogen (NiH_2) are mainly used.

Power Control and Distribution System

This system ensures the correct voltage and current is delivered to each load when it is required. Over the length of the mission, the primary and secondary power source output varies. This can be due to, for example, different amounts of sunlight reaching the solar array or by the degeneration of the battery. A shunt regulator is used to convert any excess electrical energy into heat, which is then radiated away into space through a radiating plate.

Thermal Control Systems

The control of the temperature of the equipment and structure of a spacecraft is important for two main reasons. Outside a relatively narrow temperature range, equipment will usually not operate efficiently or reliably. Secondly, thermal expansion and contraction of materials can cause distortion in the structure. As most equipment onboard a spacecraft was originally designed for use on the Earth and because it is usually easier and cheaper to get equipment developed and tested at room temperature, most spacecraft equipment also needs to be operated at about room temperature. For rechargeable batteries, the temperature can vary between about 0 °C and 20 °C; most electronic equipment operates between about –15 °C and 50 °C and mechanisms such as gyroscopes and solar array drives operate between about 0 °C and 50 °C.

The spacecraft temperature is a balance between the heat gained and the heat lost. Rocket motors, electronic devices and batteries all generate heat and add to the heat gained within the spacecraft. Heat is also gained from the outside environment. This is mainly from direct radiation from the Sun. The reflected heat or albedo radiation from the dayside of nearby planets and their constant thermal radiation also contribute to the heat gain. Heat is mainly lost by radiation into space. Before a spacecraft is launched, these thermal variations must be analysed for each part of the mission. If, as is likely, this analysis shows that some equipment will reach a temperature outside of the operating limits, the design of the spacecraft must include ways of obtaining an acceptable temperature, such as with the use of heaters or radiators or by using surfaces that reflect or absorb the Sun's heat or by moving the satellite to ensure that specific areas are subjected to either the heat from the Sun or the cooling effect from areas in shadow.

Thermal Protection for Re-entry Vehicles

The capsule that is used to bring instruments or living things safely back to the surface of the Earth is called the re-entry vehicle. In a low Earth orbit, the vehicle will be travelling at about eight kilometres per second, and so as it begins to pass through the atmosphere, it will produce frictional heat, which can reach over 2,000 °C. This is hot enough to melt steel and can cause the vehicle to glow or even burn up. This is what happens when a spacefaring particle of stone or metal known as a meteoroid enters the atmosphere. The result is a meteor or shooting star in the sky. Therefore, if we want anything to be returned to Earth safely, the vehicle has to be designed to withstand the frictional heating. The earliest satellite designers did not worry about this problem, as the spacecraft were not expected to survive re-entry. Indeed, some experts thought it was not possible for a small craft to survive re-entry, but as it was of military importance for ballistic missiles to be able to survive, a lot of research was undertaken into the problem. In the early 1950s, this research focused on long needle-like designs. However, when these were tested in wind tunnels, they burnt up from the frictional heating. In 1952, the US scientist H. Julian Allen, at the Ames Aeronautical Laboratory of the National Advisory Committee for Aeronautics (NACA), the precursor to NASA, made a counterintuitive discovery. He found that an increase in the drag of the vehicle reduced the heat generated. This is because blunt bodies with a nose that is flattened, as if a hammer has hit it, form a thick shockwave ahead of the vehicle when it travels faster than the speed of sound. This both deflects the heat and also slows the body more quickly. The shockwave can be seen in the shadowgraphs in Figure 2.4. Allen and colleagues pioneered and developed the blunt body heat shield designs, which were later used in the *Mercury*, *Gemini* and *Apollo* space capsules. A shadowgraph makes fluid flow disturbances visible by passing light through a flowing fluid. The light is refracted by the differences in the fluid and bright and dark areas

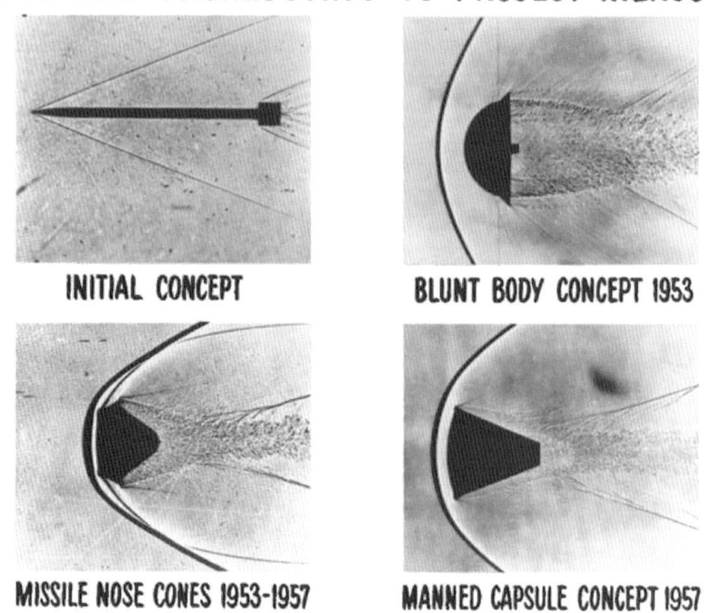

FIGURE 2.4 Blunt Body Shadowgraphs.
Image courtesy NASA

can be seen on a screen placed behind the fluid, just as the hot air rising from a fire can be seen as a shadow cast by sunlight.

Thermal Protection System

The shape of the spacecraft cannot alone deflect all of the heat from reaching the vehicle and so research was conducted on other methods of cooling the re-entry vehicle. Heat sinks were investigated, which conducted the heat away from the surface of the vehicle to a mass of material that soaked it up. The best material for this proved to be copper, and by putting a large amount of copper just below the outer shell of the vehicle, the craft could be prevented from burning up. However, the heat sinks themselves reached very high temperatures, temperatures that man or most other items inside the space capsule could not survive. Heat sinks are also very heavy and reduced the available payload mass considerably. Other ideas were investigated, including transpirational cooling and re-radiation. Transpirational cooling involved a liquid being boiled off with the change from liquid to gas taking away the heat. Re-radiation is where the heat is radiated away. Liquid metal cooling, where a liquid metal, such as mercury, was circulated through the heat shield and, therefore, conducted the heat away, was also investigated. The most promising idea though was ablation. The re-entry vehicle would be coated with a material that absorbed heat and charred. It either then flaked off or vaporized, taking the absorbed heat with it. The first *Mercury* spacecraft used a blunt body design and a heat sink on its re-entry capsule. Later versions used the blunt body design and an ablative surface. A blunt body design and improved ablative materials were used on the *Gemini* and *Apollo* spacecraft. As the ablative surface could only be used once, a different type of heat protection system was needed for the Space Shuttle. This is the Thermal Protection System, which consists of reinforced Carbon–Carbon, low- and high-temperature reusable surface insulation tiles, and insulation blankets. There are approximately 24,300 tiles and 2,300 blankets on the outside of each Space Shuttle.

3. Space Missions

Getting anything into space is expensive. The planning stage of any mission is therefore crucial, not just for the successful completion of the mission, but also to ensure that it remains within the time and financial budgets. Before any mission can be launched various stages must be completed. First the initial concept must be explored, and then the detail developed. Next the spacecraft and launch vehicle must be produced and tested and finally the pre-launch associated activities conducted. The time scale for all this depends on the complexity of the mission. Large and complex missions usually take between 10 and 15 years to develop and, once launched, they can operate for 5–15 years. Some small and less complex missions can be developed within 12–18 months and operate between six months and several years.

The initial objective of a space mission may be as vague as "to land a man on the Moon and return him safely to the Earth". The mission requirements or elements are determined by analysing the mission and its objective. These often mean that the mission must be completed at the minimum cost and risk. All space missions incorporate the seven basic elements listed below:

1. The subject, or what is being investigated, measured, or interacted with
2. The launch system, which includes the launch site, the launch vehicle, upper stages and the associated equipment both on and off the ground. The launch system determines the size, shape and mass of the spacecraft
3. Mission operations. This is the hardware, software and people that are involved in the mission, the concept, the policies and procedures and the flow of data between them
4. The spacecraft, which is made up of the spacecraft bus and the payload. It also includes all the hardware and software that is used to investigate, measure or interact with the subject and all of the equipment that is required to keep the payload in the correct orbit and attitude. These items are covered

in more detail in Chapters 2 – "Rockets and Spacecraft", 4 – "Movement in Three Dimensions", and 6 – "Navigation in Three Dimensions"

5. The orbit. This influences every element of the mission and is discussed in Chapter 4 – "Movement in Three Dimensions"
6. The communications architecture. This covers the communication, command and control of the spacecraft and is covered in more detail in Chapter 7 – "Communication"
7. The ground system. This consists of the ground stations and communication links on the Earth and allows the spacecraft to be tracked, controlled and the data and telemetry information from the mission to be distributed to the users and operators. This is also covered in more detail in Chapter 7 – "Communication"

The first stage of the mission plan is to determine whether the concept is feasible, called concept exploration. This is done by examining the mission concept, the required performance, the physical practicalities, such as the current technologies in launcher and spacecraft capabilities and the laws of physics, and the available launch sites. Other constraints include political agendas, cost and risk.

After the concept exploration an initial design for the system, called a point design, is developed. This design is capable of meeting the objectives of the mission, but it is not optimised to, for example, reduce costs, weight or improve the performance.

If the mission is feasible, the next stage is mission analysis. This term is usually used to describe the process of quantifying the objectives of the mission, how well these objectives must be met and comparing these figures with the results from various solutions. The available solutions, such as the different possible operational orbits or trajectories, different times when the launch can take place, known as the launch window, and different launch vehicles, are compared to produce the optimum mission that will meet the objectives, within the budget and risk constraints. The optimum mission is usually a compromise between the scientific objectives of the mission, the possible orbits, the launch vehicle, the spacecraft and the ground stations. There is not currently a comprehensive computer model that can take all of the objectives and possible solutions and calculate the perfect mission. Therefore analysts are required to investigate as many of the options as possible within their own time and budget constraints.

The initial analysis will determine the preliminary stages of the mission, such as the desired orbit, the launch window, the number of rocket stages, the strategies for the transfer from the initial orbit to the final orbit, including the optimum number and direction of manoeuvres and any use of planetary gravity assists or low-thrust electric propulsion systems. Gravity assists are explained in more detail in Chapter 4 – "Movement in Three Dimensions" and electric propulsion systems in Chapter 5 – "Propulsion Systems". The initial analysis will also recommend how the spacecraft is to remain in the desired orbit and what will happen at the end of the satellite's life. The options for an Earth-orbiting mission include burning up in the Earth's atmosphere, landing safely somewhere or getting moved to a graveyard orbit.

When the different aspects of the mission are specified, the detailed development can begin. This includes choosing the launch site and launch vehicle, deciding which ground stations can be used for communication and designing the actual spacecraft and instruments required. Once designed, the manufacture and testing of each component and system can be completed. Contingency plans are also made, in case of an emergency during any part of the mission. Once the spacecraft is produced and ready to be deployed, the pre-launch activities begin and the mission is then continued through to the end of the life of the spacecraft.

Launch System

Launch Sites

A rocket launch site must meet many criteria. Unless a perfect site or location is available, a compromise must be made between the mission requirements, geography, climate and logistics. The most important requirement is safety. This has led to most launch sites being built in remote, unpopulated areas, such as deserts or coastal areas, as an accident involving burning fuels in an urban area could be disastrous. The climate of the site makes a lot of safe areas such as islands and deserts unsuitable, as extremes of temperature, unpleasant working conditions, humidity and cyclones will all cause difficulties and limit when rockets can be launched.

The geography, such as the terrain, groundwater level and the type of soil, influences the design and structure of the launch facilities and therefore determines what style and how large a launch vehicle can be used at the site. The logistics include the administration and maintenance of the site and the transportation facilities for the supplies and personnel and also other facilities such as housing for the personnel to operate the site. Sites such as Cape Canaveral in Florida, USA and Kourou in French Guiana have most parts delivered by sea, whereas Baikonur in Kazakhstan and Vandenberg in California, USA have items delivered by rail. Wallops Island in Virginia, USA and Xichang in China use mainly road links.

The latitude of the launch site affects the type of orbit a rocket can be launched into. If the launch site does not lie somewhere beneath the path of the required orbit, the orbit must be changed once in space, which is costly and, because more propellant is required, it limits the mass of the payload. Rockets can be launched from the European Spaceport at Kourou into almost any orbit, as not only is it near the equator but it is also able to launch rockets at an angle of up to 102°, which makes polar orbits possible. The current Russian Federation is too far north to be able to launch a manned craft to a rendezvous with the International Space Station, so the Russians lease Baikonur from Kazakhstan.

Choosing the Launch Site

The mission requirements will usually determine which launch site and launch vehicle can be used. Most large launch vehicles often only have dedicated launch facilities at one spaceport. For example, ESA's *Ariane 5* can only launch from Kourou and the Space Shuttle usually launches from the Kennedy Space Centre in Florida. A *Soyuz* launch base is at Baikonur, but one is also currently being constructed at Kourou, which will make it possible to use the equatorial position to significantly increase the launcher's capability.

Some launch sites are more suited to certain orbits. In the USA, there are two major launch sites. For launches to polar orbit, the Western Space and Missile Centre at Vandenberg Air

Force Base in California is often used. From there, rockets are launched southwards over the Pacific Ocean into polar orbits. For launches to inclined and equatorial orbits, the Eastern Space and Missile Centre at Cape Canaveral Air Force Station in Florida and the adjacent Kennedy Space Centre are used. Rockets are generally launched due east from there, to take advantage of the Earth's spin. At the equator, the ground is travelling at about 465 metres per second from west to east. At the poles it is stationary. By launching from near the equator in an easterly direction, the ground's speed is added to that provided by the launch vehicle. Although the Earth's rotation is only about 6% of the 8,000 metres per second required to get into orbit, the fuel saving made by starting off with a speed of 465 metres per second is large.

Once a site for a launch complex has been chosen, a technical centre, a launch complex and a control centre are required. At the technical centre, the parts of the launch vehicle and the payload are assembled and checked and prepared for launch. Once the launch vehicle is prepared for flight at the launch complex, the countdown can begin at the control centre.

Sound Suppression

The sound produced at the launch of a rocket can cause damage to the launch vehicle and the payload. Some launch sites have the capability to suppress the sound. For example, Kennedy Space Centre (KSC) uses sound suppression techniques to help protect the Space Shuttle from damage during the launch. The acoustical energy levels peak when the Space Shuttle is about 91 metres above the launch platform and they cease to be a problem at an altitude of about 305 metres. The KSC system includes a tank that holds over 1.1 million litres of water. The water is released just before the ignition of the Space Shuttle engines and flows through pipes over 2 metres in diameter for about 20 seconds. The water reduces the amount of sound energy reflected off the launch pad and back towards the Shuttle. The water also prevents the reflection of the rocket exhaust damaging the Shuttle. Figure 3.1 shows a test of one of the systems at KSC. The crowd of people on the platform give some scale to the photograph.

FIGURE 3.1 Sound Suppression Test. The Water Protects the Space Shuttle During Launch.
Image courtesy NASA/KSC

Launch Windows

Other than for equatorial orbits launched from the equator, any launch site will only pass directly under the path of the orbit a maximum of twice a day. Therefore the available time period to launch into the desired orbit is limited, and is called the launch window. If the launch site does not pass under the desired orbit either the orbit will need to be adjusted after launch or an alternative launch site needs to be used. As the direction of the launch may be restricted from the launch site, there may only be one launch window available per day. The time when the rocket can be launched may be further limited by other restraints imposed by the mission requirements. For example, for certain missions, it is not just the correct orbit that is required, but also the correct time. In space, everything is moving, either in relation to each other, such as the Earth going around the Sun, or on its own axis, such as the Earth's rotation that gives us day and night. If a spacecraft is required to rendezvous with something, for example, the Space Shuttle to the International Space Station, or to fly by a planet for gravity assist, the launch must be timed so that the spacecraft arrives at a particular place at a particular time. This can limit the number of days when the mission can be launched. This is called the launch period. The launch period can be further reduced when, for example, multiple gravity assists are required. The launch period can be repeated at intervals or can be a one off opportunity. The launch window occurs each day within the launch period. For missions in low Earth orbit this launch period is 365 days a year and so it is not a problem. The length of the launch window depends on the mission and also the spacecraft used. The spacecraft *Gravity B* had a one second launch window. This was because the plane of the orbit had to be aligned with a guide star to about +/− 0.04° of longitude. As the plane of the guide star, defined by the centre of the Earth, the North Pole and the guide star, does not rotate with the Earth, the speed with which the Earth rotates limited the launch window. As the Earth rotates through 360° in almost 24 hours or 86,400 seconds, the Earth turns through about 0.004° every second. This means that, in theory, the maximum launch window was ten seconds long. However, the maximum accuracy of the launch vehicle, a *Delta II* rocket, is only about +/− 0.03°, which only left about a two second margin in the maximum theoretical window. Other considerations, including a factor of safety, had to be taken into account in case, for example, the rocket was launched a fraction of a second early or late. All that remained was the one second window. If there were any problems in the last few minutes of the countdown, the launch would have been abandoned and rescheduled for the next opportunity.

It is not only the accuracy of the launch vehicle that influences the length of the launch window. The launch vehicle will have been chosen as it can launch a certain mass at a certain velocity. If the payload is at the maximum, so that the launch vehicle can only just get it to the correct orbit under the best conditions, then the launch window becomes very narrow. However, if extra fuel is carried, at the expense of some of the payload, the thrust can be greater and the rocket can travel a bit faster or further and catch up with the original plan, therefore the launch window widens around the ideal time. However, so fuel and money are not wasted, there is a balance between the mass of the payload and the energy capabilities of the launch vehicle. For a spacecraft going into a low Earth orbit, which has no strict time requirements and a large launch period, the launch window can be shorter. However, for missions where the launch period is relatively short and will not occur again for a long time, such as a mission to Mars, a longer launch period is required to take account of adverse weather conditions or delays in the preparation of the launch vehicle. With this type of mission, lift-off is usually planned for before the ideal launch day. The rest of the launch period is made up of a range of days after the ideal launch date. If the launch period is missed, the window of opportunity for a mission to Mars closes for another two years. Once the spacecraft has been launched, it is moved into the required orbit, as described in Chapter 4 – "Movement in Three Dimensions".

Landing Sites

Earth Landing Sites

The type of spacecraft and its payload determines the type of landing site required. Yuri Gagarin, the first man in space, ejected from his spacecraft *Vostok 1* when it was in the Earth's atmosphere and parachuted to land. The *Mercury*, *Gemini* and *Apollo* astronauts stayed within their capsules, which were parachuted down into the ocean. The Space Shuttle glides in to land on a specially designed runway. Manned spacecraft need to land within easy access of recovery teams, whereas unmanned capsules can be left for a while before they are recovered. All astronauts and cosmonauts undergo extensive survival training in case no rescue party can reach them quickly and they have to rely on themselves.

Space Shuttle Landing

Since the Space Shuttle is launched from NASA's Kennedy Space Centre (KSC) in Florida, it is also the preferred landing site. This saves time and money, as landing at any other site requires the Space Shuttle to be transferred back to KSC. This is done on top of the Shuttle Carrier Aircraft, which is a modified Boeing 747, as can be seen in Figure 3.2. The preferred backup landing site is at Edwards Air Force Base in California, where the weather is more stable and

FIGURE 3.2 Space Shuttle Discovery, on top of the Shuttle Carrier Aircraft, touches down at NASA Kennedy Space Centre.
Image courtesy NASA/KSC

predictable. The weather is a major factor in whether the landing is at KSC, Edwards Air Force Base or if it is postponed until a later orbit. The weather conditions include the amount and height of any cloud cover, the visibility, the wind speed and direction and if any thunderstorms are in the vicinity. The angle of the Sun is also considered, in case it is in the pilot's eyes as they come in to land. The chosen landing site can be changed up to 90 minutes before landing. About an hour before landing a de-orbit burn slows the Space Shuttle enough to begin its descent. There are other emergency landing sites around the world, which are covered later in this chapter.

The Shuttle Landing Facility (SLF) at KSC is shown in Figure 3.3. It was designed specifically for the returning Space Shuttle. It is over 4,500 metres long and about 90 metres wide, which is longer and wider than those at most commercial airports. In comparison, London Heathrow's longest runway is just over 3,900 metres long and only 45 metres wide.

The SLF runway is made of 40 centimetres thick concrete and slopes gently from the centre to the edges to help drainage. Although it is only a single landing strip, it is considered to be two runways as the Space Shuttle could approach from either the northwest or the

FIGURE 3.3 Space Shuttle Landing Facility at Kennedy Space Centre. Image courtesy NASA/KSC

southeast. At the southeastern end of the runway there is a parking apron or ramp that is about 170 by 150 metres, which incorporates a Mate/Demate Device (MDD) capable of lifting over 100,000 kilograms. The MDD is used to lift the Space Shuttle on or off the Shuttle Carrier Aircraft when it needs to be ferried around.

The Edwards Air Force Base has several dry lakebed runways and one hard surface runway. The longest strip, which is part of the 114 square kilometre Rogers Dry Lake, is just over 12 kilometres long, although a concrete runway is preferred for night landings as the dust from the lake bed can obscure the lighting. As the Space Shuttle is only the size of a DC-9 jetliner such large runways are not essential for landing. However, unlike a normal plane, the Space Shuttle is essentially a glider during its unpowered re-entry and landing and, as it does not have the power to go around and make a second try, each landing must be perfect the first time. The touchdown speed of the Space Shuttle is around 350 kilometres per hour, whereas a DC-9 usually has a touchdown speed of about 240 kilometres per hour. Larger runways are therefore used to help increase the safety margins of landings and a 12 metre drag chute is used to help the Space Shuttle roll to a stop. After landing, the Space Shuttle crew and about 150 other personnel make the Space Shuttle safe and into a suitable mode for towing to the Orbiter Processing Facility. After the Space Shuttle has cooled down and the crew have departed, which usually takes less than an hour, the responsibility for the Space Shuttle is handed from the Johnson Space Centre in Houston, Texas to the Kennedy Space Centre in Florida, who are then responsible for the turnaround of the Space Shuttle and preparing it for re-launch.

Any material on the runway that should not be there is classed as foreign object debris and is a potential hazard. Workers continually check the runway up to about 15 minutes prior to landing. Birds are also a hazard and are of particular concern at KSC as it is a national wildlife refuge. Special pyrotechnic and noise-making devices, as well as selective grass cutting, are used to discourage the birds from the area around the runway.

The control tower controls all of the local flights and ground traffic. This includes the support aircraft for the Space Shuttle launch and landing, helicopters for security, medical evacuation and rescue and NASA weather assessment aircraft. Near the control tower there is a fire station and a viewing area for the press and guests and also the Shuttle recovery convoy staging area for the recovery team. This holds specially designed vehicles or units, which help to make safe the Shuttle, assist in the crew's departure and also the tractor to tow the Shuttle to the processing facilities.

Other Manned Earth Landing Sites

Most manned capsules that were launched by the former Soviet Union, and that are now launched by the Russian Federation, land within the borders of Kazakhstan, on the vast region of open plains of the Kazakh Steppes. Ground control stations in both Moscow and Baikonur follow the touch down of the *Soyuz* capsule. Once they have landed, the crew deploy one or more communication antennas so the recovery teams can find their precise location. The accuracy of the planned landing is usually within a range of 30 kilometres, in one of two areas in northern Kazakhstan, one near the town of Arkalyk, the other near the town of Dzhezkazgan. However, in case of emergencies, *Soyuz* spacecraft can land anywhere in the world, including on water. On April 22, 1968, the United Nations initiated an agreement for just such a contingency for any spacefaring nation, called the "Agreement on the Rescue of Astronauts, the Return of Astronauts and the Return of Objects Launched into Outer Space" It states that:

> If, owing to accident, distress, emergency or unintended landing, the personnel of a spacecraft land in territory under the jurisdiction of a Contracting Party or have been found on the high seas or in any other place not under the jurisdiction of any State, they shall be safely and promptly returned to representatives of the launching authority.

In Kazakhstan the recovery teams reach the landing site by helicopter as soon as possible after the spacecraft touches down. The hatch is opened and the crew are assisted out. The crew then sit in chairs while answering questions from the recovery team. A medical tent is prepared on site for initial medical checks on the crew. Other recovery team members cordon off the area and gather the landing parachutes. After the medical checks, the crew are taken by helicopter first to Astana, the capital of Kazakhstan and then back to Star City, to the northeast of Moscow. At Star City they undergo more medical tests and stay in quarantine for two weeks while they readapt to life back on Earth and also evaluate their mission.

The Chinese manned missions touch down on the grasslands of Inner Mongolia in northern China and are recovered in a similar manner to the crews landing in Kazakhstan.

Before NASA used the Space Shuttle, their astronauts landed on water. This system used the water to cushion the spacecraft as it landed and extra braking rockets were not required, as they are on the Russian and Chinese manned spacecraft. This type of landing is known as splashdown. On the early *Mercury* flights, the spacecraft was hoisted by helicopter to a nearby ship. After the capsule *Liberty Bell* 7 sank, this method was changed so that all later capsules had a flotation collar attached to the spacecraft, which increased the buoyancy. The spacecraft was then brought alongside a ship and lifted onto the deck by crane. The crew could either stay in the spacecraft or climb out of the hatch and be hoisted into a helicopter to ride to the recovery ship.

Unmanned Earth Landing

An unmanned spacecraft landing site still needs to be accessible, as the recovery of the payload is important. Since Kazakhstan has become an independent state, the Russian Federation has designated another landing area for unmanned spacecraft within the borders of Russia. This reduces the extra costs involved in transporting items from Kazakhstan. NASA usually uses the US Air Force's Utah Test and Training Range to recover returning capsules. NASA's *Stardust* return capsule, which brought back samples from comet Wild 2, landed by parachute inside a landing zone of 44 by 76 kilometres. This was large enough to allow for the parachute being taken in any direction by the wind. NASA's *Genesis* mission that collected particles of solar wind, contained delicate equipment and so landing by parachute was not a suitable option. Instead, a mid-air recovery was planned. The capsule was to be slowed by a parachute and then captured by a helicopter before it hit the Earth. Unfortunately, although the capsule did land within the projected path, the parachute did not open and it hit the ground at over 300 kilometres per hour.

Landing on Other Bodies in the Solar System

Landing sites on other bodies are mainly determined by the mission requirements and also by any engineering constraints. The environmental conditions, such as slopes, abundance of rocks, altitude and temperature, must be suitable for a safe landing but the site must also be a scientifically interesting place. Extremes of weather conditions, such as are found at the poles, are usually avoided. The time of day of the landing will be important. If there is a descent imager the terrain needs to be sufficiently illuminated and if the lander is powered by solar arrays it must be able to receive enough sunlight. The number and length of available relay links for data transfer between the lander and the orbiter need to be considered from each potential landing location. If a direct Earth link is needed, the direct view to Earth may be constrained by seasons.

Before a spacecraft lands on another body in the solar system, it first orbits it. The inclination of the orbit restricts the landing areas, although if sufficient fuel has been carried, the orbit can be changed. Any atmosphere on the body will also influence the landing site as winds can affect where the lander finally comes to rest. The atmosphere or lack of one will also determine the method of landing, such as using aerobraking and parachutes or relying on retrorocket thrusters. Most landers are bespoke and are designed for certain mission and scientific requirements. They are also designed to withstand the environmental conditions they will encounter.

Moon Landers

The first spacecraft to make a soft landing on the Moon, rather than crash onto it, was the Soviet Union's *Luna 9* in 1966. It landed in Oceanus Procellarum or the Ocean of Storms. About five months later the American spacecraft *Surveyor 1* also made a soft landing on the Moon. It landed in a flat area inside a 100 kilometre diameter crater to the north of Flamsteed crater in the southwest of Oceanus Procellarum. When the spacecraft reached an altitude of about 75 kilometres and a velocity of just over 2,600 metres per second, the main retrorockets fired and, after slowing the spacecraft to about 110 metres per second and at an altitude of about 11 kilometres, they were then jettisoned. Small rocket engines continued to slow the descent until it was about 3.4 metres above the surface after which the lander fell freely under the pull of the Moon's gravity. Both the Soviet and American lunar missions in the 1960s and 1970s were used to gather information about the Moon both for scientific purposes and also for the planning of possible future missions, including manned missions. The main landing site criteria were therefore similar to those for the manned Moon landings, discussed below. Between 1976 and 1990 there were no missions to the Moon. In 1990 Japan's *Hiten* spacecraft first flew by, then orbited and then impacted on the Moon three years later. The primary reason for this mission was to test and verify technologies for future lunar and planetary missions. There has not been a soft landing on the Moon since 1976.

Manned Moon Landings

The first landing site for a manned craft on the Moon was determined mainly by safety and operational criteria. Any scientific investigation was a secondary consideration although this did become more important in later missions when more was known about the practicalities of a Moon landing. The most important safety rule stated that the spacecraft must be on a free-return trajectory. This meant that if the main engine failed, and the spacecraft could not be put into an orbit around the Moon, it would swing around the Moon under the influence of the Moon's gravity and head back towards the Earth. *Apollo 13* used this free-return trajectory after an explosion onboard forced the landing mission to be abandoned. A free-return trajectory places the spacecraft in the equatorial region of the Moon and so the landing site had to be within a belt 5° north and 5° south of the Moon's equator.

The timing of the first *Apollo* landing attempts was also important. The lunar module crew needed to view the landing area and choose a safe landing site. Therefore, the landing had to be not too long after lunar sunrise, when the Sun's height above the horizon was enough to highlight the surface, without producing long and confusing shadows, but not too high as to wash out all of the details. The launch from the Earth was chosen so that the lunar module would land when the solar illumination was near optimum. However, the launch time was constrained to daylight hours at the launch site, in case of an aborted launch and an emergency rescue operation was required.

The angle of the Sun was also relevant after the lunar module had landed. As the Moon has no atmosphere, sunlight is not scattered as it is on the Earth and the shadows are completely black. Therefore if the Sun were too low, visual observations would have been difficult. If the Sun were too high, there would be no shadows for contrast, and again visual observations would be difficult. The temperature on the Moon also varies with the angle of the Sun. To protect the astronauts and the spacecraft, the landing site was specified to be when the Sun was between 15° and 45° above the horizon.

If the launch were cancelled for any reason, it would take nearly 48 hours for the *Saturn V* launch vehicle to be ready for use again. Although a day on the Moon lasts just over 27 Earth days, and the angle of the Sun over the horizon changes slowly, a delay of 48 hours would mean that the original landing site would be washed in sunlight and the fine detail shown up by the shadows would not be visible to the lunar module crew. Therefore a backup landing site, that was more westerly than the original site, was needed. The launch period, which was just a few days per month for one landing site, was substantially increased when an alternate landing site could be used.

The favoured landing sites were therefore on the eastern side of the Moon's visible face. The east and west boundaries were limited by navigational and communication constraints. Earth-based computers were used to calculate both the lunar module's orbit and any course corrections necessary, from data provided from the lunar module. When the spacecraft was behind the Moon it was out of radio and radar contact with the Earth until it reappeared around the eastern edge. The time taken to send the data back to Earth and to receive and act on any corrections limited the eastern boundary of the landing area to 40° E. Another factor limiting both the eastern and western boundaries was the lack of accurate maps of the areas. At the time, some visible surface features were up to 1,800 metres different from where the best maps placed them and the most detailed telescopic view could only distinguish items about the size of a football field. However, the maps of the central part of the visible face of the Moon were more accurate. The western boundary was also the 40° mark and this gave a strip of lunar terrain approximately 300 kilometres wide centred on the lunar equator and about 2,400 kilometres long, which became known as the *Apollo* Landing Zone. It extended from the southeastern edge of Mare Tranquillitatis, which is close to where *Apollo 11* eventually landed, to a point northeast of Flamsteed crater in Oceanus Procellarum.

After the landing zone was defined, the terrain and geology of the Moon's surface determined the potential landing sites. As the telescopic views from Earth were not good enough to provide this detail, the Ranger series of unmanned missions were used to obtain the close-up images of the surface. These spacecraft were designed to collide with the Moon's surface and return imagery until they were destroyed upon impact. The astronomers of the time held numerous views about the physical nature of the Moon's surface. Some believed the smooth looking areas on the Moon may be covered with a fine dust several metres thick, and so a lunar module may sink beneath the surface. Others thought that the lander might not be able to find a level surface on which to land, as it may be cluttered with rocks and boulders and pitted with small craters. These thoughts lead to the next unmanned missions, the landers of the *Surveyor* program, which proved the viability of a soft landing on the Moon. Instruments carried onboard were used to help evaluate whether their landing sites were suitable for the manned *Apollo* landings. Some *Surveyor* spacecraft even had robotic shovels that were designed to test the physical characteristics of the lunar soil. The landing site had to be relatively flat and provide a large expanse of smooth surface to allow for errors when descending.

By the time the manned mission *Apollo 10* was launched, the prime landing area had been selected. The spacecraft made a low pass over it so the crew could check its suitability. Although there was a rough area at the end of the site, it was chosen as the landing area for *Apollo 11* and the first men on the Moon. Five other *Apollo* missions landed men on the Moon between 1969 and 1972, all within the *Apollo* Landing Zone.

Planetary Landing

In 1970, a landing capsule from the Soviet spacecraft *Venera 7* landed on Venus. It was the first spacecraft to return data from the surface of another planet. While it was parachuting through the atmosphere it returned data signals for 35 minutes and, after landing, weak signals were received from it for a further 23 minutes. Hardly any information was available about the surface of Venus at the time, and so any form of landing was rated as a success. The harsh conditions of high temperatures and pressures on the surface of the planet and the high winds and sulphuric acid and compounds of chlorine and fluorine in the atmosphere caused the short life of the capsule after it landed. *Venera 8*, which was launched in 1972, confirmed *Venera 7*'s findings. The survival of the lander was expected to be limited and so the landing of *Venera 8* had to take place on the day-lit side of the planet as it also measured the amount of light on the surface. This showed that surface photography would be feasible, as there was a similar amount of light to Earth on an overcast day with roughly 1 kilometre visibility. The following spacecraft to visit Venus all landed on the day-lit side, so photographs could be taken and transmitted back to the Earth.

The first successful Mars landing was NASA's *Viking 1*, which landed in 1976. This time, the orbiter imaged the surface of the planet to find a safe landing site for the lander. The initial landing site was too rocky, and so after another landing site was chosen, the lander separated from the orbiter and landed in Chryse Planitia at 22.48° N, 49.97° W. *Viking 2* landed on the opposite side of Mars and nearer the North Pole. Mars' atmosphere is insufficient to rely on parachutes alone to slow the descent so retrorockets are also usually used. However, the atmosphere is thick enough to produce frictional heating and winds, which need to be considered when designing a lander.

ESA's ill-fated *Beagle 2* mission was designed to use large airbags that were to inflate around the lander and protect it when it hit the surface instead of using retrorockets. Its mission was to land on Isidis Planitia and search for traces of life. Isidis Planitia is a large flat sedimentary basin where traces of life could have been preserved. The landing site was chosen to be at a low latitude, which would be warmer than the higher latitudes, and therefore minimized the amount, and therefore mass, of thermal protection needed to protect it from the cold Martian night. The site is also low lying, which meant the parachutes had longer to work and slow the descent. Images taken by the orbiter *Mars Express* and other missions to Mars showed that the surface within the landing area was probably hardened dust with about an eighth of the area covered by rock fragments, mostly smaller than house bricks in size. Therefore it was expected to be not too rocky to threaten a safe landing, but rocky enough to be interesting for the experiments and also not too dusty. *Beagle 2* should have landed on December 25, 2003. Unfortunately, no communication was ever received from it, and it was declared lost in February 2004.

Other Landings

ESA's *Huygens* probe left the *Cassini* spacecraft and landed on Saturn's moon Titan in 2005. As Titan has an atmosphere, aerobraking and three sets of parachutes were used to slow the probe's descent. The probe descended during daylight hours so photographs could be taken during the descent and also after it had landed.

NASA's *Near Earth Asteroid Rendezvous NEAR/Shoemaker* mission to the asteroid 433 Eros completed all of its scientific goals while in orbit around the asteroid in 2001. The orbiter was then made to descend and to take close-up pictures of the surface of the asteroid. Although the orbiter was not designed to land it came to rest on the tips of two of its solar panels and the bottom edge of its body and continued to operate and communicate with Earth, sending back valuable engineering and scientific data. This was the first soft landing on an asteroid. The touchdown speed was between 1.5 and 1.8 metres per second, which was one of the softest planetary landings ever. The asteroid has no atmosphere, so thrusters were used to slow the orbiter during four different braking manoeuvres. The irregular shape of Eros required that, when in a very low orbit, the *NEAR/Shoemaker* spacecraft orbited the asteroid in the opposite direction to which the asteroid was spinning. If it had orbited the same way, the unevenness of Eros' gravity field may have caused the spacecraft to be ejected from its orbit or pulled in to crash on the asteroid's surface.

Japan's *Hayabusa* mission was the first mission designed specifically to land on an asteroid. As mentioned previously, it landed on asteroid Itokawa in 2005. The mission was to take a sample of the asteroid and return it to Earth. As the size, shape and spin-rate of the asteroid were not known very accurately before the launch, the *Hayabusa* spacecraft needed autonomous navigation, guidance and control systems to gather topographic and range information about the asteroid's surface, which it sent back to the team on Earth. The team then decided where the lander should try to land. There were some technical problems and the separate landing vehicle or rover, called *MINERVA*, which is short for MIcro/Nano Experimental Robot Vehicle for Asteroid, failed to land. However, the spacecraft itself managed to land and lift off twice, before heading back to Earth. It is hoped that it acquired some samples of the asteroid, although this will not be known until the spacecraft returns to Earth in about 2010. *Hayabusa* used the Earth for a gravity assist and also a solar powered electric propulsion system to reach the asteroid and an electric propulsion system to land and lift-off from the asteroid.

ESA's *Rosetta* mission, which was launched in 2004, aims to land a probe on comet 67P Churyumov-Gerasimenko in 2014. To reach the comet it will use three Earth and one Mars gravity assist manoeuvres to increase its speed. After the spacecraft has entered into a one kilometre high orbit around the comet's nucleus and has mapped it, a lander, named *Philae*, will be deployed and attempt to complete the first ever soft landing on a comet. *Philae* will approach the comet very slowly, using a cold gas thruster to descend accurately. As the comet only has a weak gravitational pull, *Philae* should land softly. However, to stop it from bouncing back off into space, a harpoon will be fired at the moment of touch-down and ice screws on each leg will secure *Philae* to the comet. It will then take scientific measurements and pictures, which will be the most detailed data from a comet ever taken.

Emergency Systems

Manned spacecraft require an emergency escape system from the launch vehicle in case of, for example, a launch pad emergency such as a fire. There are different designs, depending on the type of launch vehicle and how many crew are onboard.

Launch Escape System

A Launch Escape System (LES) uses rockets to separate and remove the crew module from the launch rocket quickly. The system must be able to take the crew to a height sufficient that when the parachutes are deployed, their module can then be slowed and land safely. To obtain this amount of thrust the system usually uses solid rocket motors mounted on top of the crew module. The rocket nozzles are pointed away from the crew module as the hot exhausts could endanger the crew. A LES is only used if the crew are in imminent danger, such as an impending explosion. The American *Mercury* and *Apollo* spacecraft used a Launch Escape System. This type of system is still used on the Russian *Soyuz* spacecraft and the UK's Starchaser Industries are developing one for use on their reusable launch vehicle *Thunderstar/Starchaser 5A* for sub-orbital space tourism.

In 1983, the *Soyuz T-10* mission caught fire on the launch pad. Fortunately, the LES was able to carry the crew capsule clear, seconds before the rocket exploded. This has been the only emergency use of a LES. The Soviet *Vostok* and American *Gemini* spacecraft both made use of ejection seats. The ESA's *Hermes* and the Soviet *Buran* space shuttles would also have made use of them if they had ever flown with crews. If possible, spacecraft designers prefer to use ejection seats as they are lighter and would be available for use when the spacecraft is returning to Earth. However, ejection seats are not practical for spacecraft with a large crew as a separate seat and exit hatch must be provided for each member.

Shuttle Emergency Egress System

The Space Shuttle had ejection seats for the test flights, but when it was deemed operational, these were removed. If there is a problem and the Shuttle is still on the ground waiting for lift-off, the crew can leave through the Emergency Egress System. This system is at the same level as the Orbiter Access Arm, which is used for access to the Shuttle crew compartment. It includes seven baskets suspended from seven wires that extend to a landing zone 366 metres away. After the astronauts climb into the baskets they slide down the wires and are stopped by a braking system of a catch net and a drag chain so they do not hit the ground too hard. The angle of the slidewire causes the baskets to move the astronauts as far away from the Space Shuttle as possible, as quickly as possible.

Shuttle Launch Abort

There are five types of abort after lift-off for the Space Shuttle. The four intact aborts are designed to provide a safe return of the Space Shuttle to a planned landing site. There is also a contingency abort. The contingency abort is only used in extreme emergency, and is designed for the crew's survival when an intact abort is not possible. The crew would then use the in-flight crew escape system, described later. A contingency abort would generally result in a ditch operation and is never chosen if another abort option exists.

The four types of intact abort are return to launch site (RTLS), transatlantic abort landing (TAL), abort once around (AOA) and abort to orbit (ATO). Which abort mode is selected depends on the cause and timing of the failure leading to the abort and which mode is safest or improves mission success.

If the Space Shuttle loses thrust from a main engine between launch and about four minutes and 20 seconds into the flight, the return to launch site abort will be used. This system will also be used if there is a failure in a major system within this time frame, for example, a large cabin pressure leak or a cooling system failure. The RTLS allows the Shuttle, crew and payload to return to the launch site approximately 25 minutes after lift-off. After four minutes and 20 seconds, there is not enough fuel to turn the Shuttle around and return to the launch site and, if it is imperative to land as quickly as possible, the transatlantic abort landing will be used.

In a TAL, the Shuttle continues on a ballistic trajectory across the Atlantic Ocean and lands on one of three runways in Europe. These runways have the necessary runway length, weather conditions and are within gliding range of a Space Shuttle trying to reach the International Space Station. There are two runways in Spain, one at Moron the other at Zaragoza, and one in the south of France at Istres. By burning the propellants through both the orbital manoeuvring system engines and the reaction control system engines, the weight of the Shuttle is decreased and the vehicle performance is increased, as this moves the centre of gravity and allows proper vehicle control. Landing occurs approximately 45 minutes after launch.

If it is impossible to achieve a viable orbit, or not enough propellant is available to place the Shuttle into orbit and then land, an abort once around is used, when the Shuttle circles the Earth once, and then comes into land. The AOA can also be used when it is necessary to land quickly, but if it is too late for a RTLS or TAL. The landing sites available in an AOA are White Sands in New Mexico, Edwards Air Force Base or the Kennedy Space Centre. The landing will be about 90 minutes after lift-off.

If it is impossible to reach the planned orbital altitude, but still possible to enter an orbit, the abort to orbit can be used. This boosts the Shuttle to a safe orbital altitude. The orbital manoeuvring system engines are then used to place the Shuttle into a circular orbit. This temporary orbit allows time to evaluate the problems and decide whether to make an early landing or to raise the orbit and continue the mission.

If the problem is a system failure that jeopardizes the vehicle, the abort mode that results in the earliest vehicle landing is chosen. This depends on when the failure occurred. As more precise knowledge of the Shuttle's position can be obtained from the Mission Control Centre at Houston than by the crew from onboard systems, Mission Control is responsible for issuing the abort command. Mission Control informs the crew periodically which abort mode is or is not available. If ground communications are lost, onboard methods such as cue cards and dedicated displays are used to determine which abort mode to use.

If the astronauts need to escape from the Shuttle immediately it has landed, they can use the emergency egress slide. This is like the emergency slide on commercial aeroplanes. The slide allows the astronauts to safely exit within one minute of the hatch opening. A secondary emergency egress can also be used. This allows the astronauts to escape through a window and they can then lower themselves to the ground over the side of the Shuttle. To quickly remove the window, pyrotechnic firing circuits within the window frame are used.

In-flight Crew Escape System

If there is a problem inside the Shuttle after it has launched, or more than one engine fails and it is not possible to land safely at an emergency location, then the in-flight crew escape system would be used. This can only be used if the Shuttle is in a controlled glide. At an altitude of about 9,150 metres a member of the crew would open a valve that equalises the cabin pressure with the outside air pressure. The side hatch is then released, by which time the Shuttle would be at about 7,620 metres. Two telescoping sections of an aluminium and steel escape pole deploy through the hatch and each crew member hooks a strap onto the pole and jumps out the hatch opening, sliding down the 3.1 metre pole. At the end of the pole they are in freefall until their parachute opens and they fall back to Earth. It takes about 90 seconds for a crew of eight to bail out of the Space Shuttle.

Mission Operations

The integration of the people, hardware, software, the mission, the policies and procedures and the flow of data between them all are called the mission operations system. This system starts with the concept exploration and continues right up to the end of life of a space mission and includes every activity such as the design, the staff training, the cost estimates, the pre-launch and the launch. The Mission Operations Plan (MOP) describes the ground based and flight operations in terms that the users and operators can use.

Testing and Pre-launch

As spacecraft are expensive and cannot usually be fixed once in space, they are thoroughly tested before launch. Test centres, such as the European Space Research and Technology Centre (ESTEC) in Noordwijk, The Netherlands, allow spacecraft to be subjected to many of the physical conditions that they will endure, for example, by simulation of the severe vibrations experienced during the launch and exposure to the noise a launch vehicle generates, to make sure that they will survive the launch. The satellites are also tested to ensure they can withstand the environment in space, such as exposure to vacuum and the extremes of temperatures. These tests can continue for months, until the test engineers are satisfied that the spacecraft is capable of performing well for the duration of its planned lifetime.

The pre-launch phase includes every operation that is required until the launch vehicle is ignited and separation has occurred between the ground launch facilities and the launch vehicle. This phase usually begins a few weeks before the launch and includes the loading of the experiments into the payload, function tests and mission simulations, as well as filling the propellant tanks.

Launch

The pre-launch ends when the countdown begins. When all personnel are cleared from the launch pad and the final checks have been made, the rocket is cleared for launch. If there are no aborts, at "T minus zero", the first stage of the launch vehicle ignites and the rocket begins its journey into space. The pre-launch and launch is usually controlled by the launch director from the launch control centre, which is situated at the launch site. The control of the spacecraft is then passed to either the Mission Control Centre (MCC) for manned flights or the control centre for unmanned flights. For the Shuttle, this happens as soon as the launch vehicle leaves the launch tower. The MCC, located at the Johnson Space Centre, Houston, Texas, then maintains responsibility for the mission until its end and the Shuttle has landed and cooled down. Other jobs are completed at other centres. The tracking and data acquisition are the responsibility of the Goddard Space Flight Centre in Greenbelt, Maryland. This involves integrating and coordinating all of the worldwide NASA and Department of Defence tracking facilities needed to support the Space Shuttle missions.

For unmanned spacecraft, control is usually handed over when the spacecraft separates from the launch vehicle. The operations teams then activate, monitor and control the various subsystems of the satellite, such as the deployment of solar panels, and undertake orbit and attitude manoeuvres. This phase is known as the launch and early orbit phase or LEOP. When the spacecraft is in its final orbit or trajectory and has been tested and functions properly, the mission objectives can begin to be met. Depending on the mission, the control is usually handed over to the satellite owner or operator after this phase. For example, ESA's Space Operations Centre, ESOC, controlled the weather satellite *MetOp* during the LEOP. When the satellite was properly configured, in its nominal orbit and ready to go, control was handed over to EUMETSAT who took over responsibility for the satellite and operated it for the duration of the routine mission.

End of Life

Once the useful life of a satellite has ended, it should be disposed of responsibly. If it is left in its original orbit it will become another piece of debris that clutters space and will become a potential collision hazard for working satellites. In the belt used for geosynchronous orbits there are only a limited number of positions for satellites and so removal of dead spacecraft from these slots is becoming ever more important. Satellites in a geosynchronous orbit are usually moved into a graveyard orbit, where they will be out of the way and reduce the risk of a collision that would generate more space debris. Satellites in a low Earth orbit are usually moved so that they will re-enter the Earth's atmosphere within a reasonably short time period. When they re-enter, they must burn up completely in the atmosphere before they reach the Earth, break up into harmless sized pieces or the re-entry must be controlled so any pieces that do land on the Earth do not cause any damage to people or property. When the US space laboratory, *Skylab* fell to the Earth in 1979, it scattered debris not only across the southern Indian Ocean but also across sparsely populated areas in Western Australia. NASA was fined AU$400 for littering in Australia. This was not paid however, as the ticket was issued in "fun". J.M. Jones, the public affairs office for the Marshall Space flight Centre at the time, accompanied the *Skylab* investigation team on their trip to Australia. His personal report on the trip was published in the Johnson Space Centre Newsletter *Roundup* on August 10, 1979 (Vol. 19, No 16). He said:

> Upon our arrival, the president of the shire (county) had arranged a mock ceremony in which an officer of the parks service ticketed NASA for littering, the evidence having been found all about the country-side.

The other option at a satellite's end of life is to retrieve it, either to repair or refurbish it, recycle it or to recover material that would be dangerous if it were dispersed into the atmosphere, such as radioactive material. The first satellites to be retrieved were *Palapa B-2* and *Westar 6* in 1984. Due to a failure of their propulsion systems, the satellites had been unable to reach a geosynchronous orbit and so astronauts Dale A. Gardner and Joseph P. Allen IV retrieved them during an extravehicular activity (EVA) period from the Space Shuttle. *Palapa B-2* was refurbished and relaunched in 1990.

Reliability

The probability that the spacecraft will successfully perform its mission for a specified time is called the reliability of the spacecraft.

This depends on, for example, the correct selection of components, the correct definition of the environmental conditions and the correct component manufacturing procedures. Often, a large proportion of the cost of a space program is spent on overcoming the effects of unreliability. The mission may be lost if anything goes wrong with critical components such as the launch vehicle, the guidance and control system, the payload components or if there are significant human errors in the design or operation. The *Hubble Space Telescope* was designed to be astronaut-friendly so servicing operations could be performed during EVA's from the Space Shuttle. These operations have extended its operating life by the replacement of aging hardware, and also enhanced its scientific capability by correcting the flaw in the primary mirror and by the installation of further instruments.

A prediction of the reliability can be made based on part failure rates and the amount of redundancy incorporated into the design. Once a spacecraft is in space, it is very difficult to perform repairs or maintenance on it. Therefore, for critical systems, a form of redundancy is used. This means that there are two or more ways of accomplishing a task, where one alone would be sufficient as long as it did not fail. The two main types of redundancy are standby and active. Standby redundancy has a malfunction detection capability that switches the reserve component or system on in the event of a primary unit failure. The spare is not switched on until it is required, which should prolong its life and reduce its failure rate compared to an energized unit. Active redundancy has all of the components operational during the complete mission. If one unit fails, the others remain capable of keeping the system operational and no detection system is required. As some missions require multiple satellites to work together, complete spare satellites are sometimes in orbit, ready for use if another one fails. The *Iridium* phone constellation of satellites has multiple in-orbit spares.

The reliability of a spacecraft may be improved by using fewer parts, higher quality parts, redundancy or by improving the environment where the systems will work, such as protecting it against high radiation levels and high temperatures. However, there is a trade-off between the cost and the mass with the level of reliability that can be accepted.

4. Movement in Three Dimensions

On the Earth we usually only move in two directions, forwards and backwards or left and right. These directions are known as degrees of freedom. Unless we have access to some form of aircraft, we have a limited third degree of freedom, the ability to move up and down. In space, everything has six degrees of freedom. There are the three degrees of freedom in a straight line, left, right, up, down, forwards and backwards, but things can also rotate or spin. One end of a spacecraft can rotate up or down with respect to the other end, this is known as pitch and can cause the spacecraft to spin like a cartwheel. The spacecraft can also rotate along its long axis like a rolling pin, which is known as roll, or around its vertical axis like a merry-go-round, known as yaw. These degrees of freedom are shown in Figure 4.1. On the Earth, even in a vehicle, these degrees of freedom are usually restricted, such as by the wheels only turning on one axis and by gravity keeping the wheels in contact with the ground.

Moving around on the Earth is relatively simple. In a car, for example, to change the speed you either accelerate or brake, and to change direction you turn the steering wheel. You are normally in complete control of where the car goes by using the friction between the tyres and the ground. If you are travelling on ice, however, and make a sharp turn, the car is likely to carry on in its original direction. Travelling in space is like being in a three-dimensional ice rink. Even if the direction the spacecraft is pointing, called its attitude, is changed, it will still carry on following the path it was on. Over short distances, this appears to be a straight line, but over a longer time, the path curves as it is pulled by gravity towards the Earth, Moon, Sun and planets. On the Earth the effect of gravity is usually ignored as it keeps us on the ground, although we do notice it when we climb a hill or if we fall over. In space, however, the force of gravity changes with the distance away from the body the spacecraft is orbiting and must be taken into account.

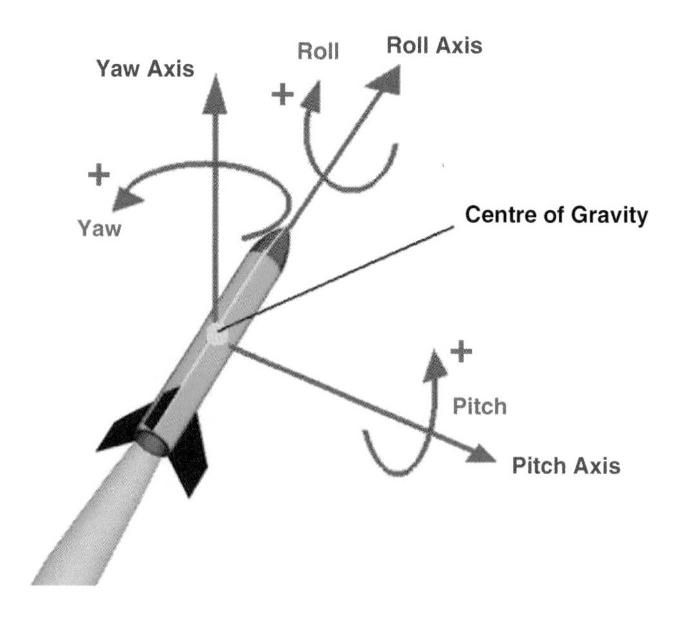

FIGURE 4.1 Degrees of Freedom.
Image courtesy NASA

Orbits

An orbit is the path that an object makes around another body, such as the Earth, another planet, a moon or the Sun. Objects in an orbit are continuously being pulled towards the centre of the body they are orbiting, as they have not escaped its gravity. If the gravity were somehow switched off, they would continue to travel in a straight line and off into space. However, the combination of the forward motion of the satellite and the force of gravity makes the path of the satellite circle the planet, as can be seen in Figure 4.2, where a satellite is in orbit around a planet. This is similar to a toy yoyo on the end of a string. If the yoyo is swung around your head, it makes big circles. The string is like the gravity, holding the yoyo in an orbit around your head. If you let the string go, the yoyo would fly off. If the yoyo stops, it falls to the ground. The speed with which a satellite orbits an object is called the orbital velocity. A satellite will orbit at a distance where its speed balances the gravitational pull, and at the same

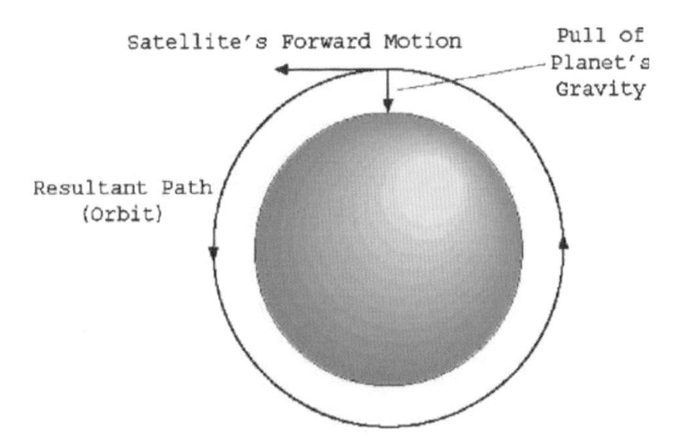

FIGURE 4.2 Forward Motion and Gravity Cause a Spacecraft to Stay in Orbit. Image courtesy NASA

point in each orbit it will be at the same height above the Earth. As mentioned before, gravity diminishes with distance, and the further something is away from the body it is orbiting, the slower it must travel to remain in orbit. If the speed changes, the size or shape of the orbit will also change.

Orbit Shape

Objects that repeatedly travel around a heavier body, such as a planet or star, have a closed orbit that they follow time and again. If the orbit is open the object just flies past the celestial body in an open curve and back off into space. Open orbits extend to infinity and are called hyperbolic. Hyperbolic orbits are used at the beginning of interplanetary missions by spacecraft that have enough speed to escape the gravitational pull of nearby planets or other objects, called the escape velocity. Most spacecraft and comets are in closed orbits around the Earth or the Sun, although some are in open orbits, such as *Voyager 1* and *2*.

In a circular orbit, the satellite will travel at the same speed wherever it is on its path, but orbits are not very often perfectly circular. The Earth's orbit around the Sun and the Moon's orbit around the Earth are both slightly squashed circles, called ellipses. An ellipse is a special type of oval shape. If a cone, like an ice cream cornet, is cut without cutting through the open end, the resulting shape will be an ellipse, as can be seen in Figure 4.3.

An ellipse can be drawn with the help of two drawing pins and a loop of string. Loop the string around the pins and put the pen inside the loop. Now pull the string taut, thus making a triangle. Move the pen around the drawing pins, while keeping the string taut. The resulting drawing will be an ellipse. Each pin is at a focus of the ellipse. The long axis of an ellipse is called the major axis, the shorter one, the minor axis. In a circle, both the major and minor axes are the same and are called the diameter, and both the foci are at the centre of the circle. Half of the major axis is called the semi-major axis. This is the equivalent of the radius in a circle. The length of the semi-major axis is used to describe the size of the ellipse. The properties of an elliptical orbit around the Earth are shown in Figure 4.4.

A circle is just a special type of ellipse and so for the purposes of orbits, it can usually be treated the same as an elliptical orbit. A satellite on an elliptical orbit around the Earth is said to be at apogee when it is furthest away from the centre of the Earth and perigee when it is closest to it. However, if something is orbiting the Sun it is said to be at aphelion when it is furthest away and at perihelion when it is closest. The generic terms for these are the apoapsis and the periapsis. For an orbit around Jupiter, they are apojove and perijove, around the Moon, apselene and periselene or apolune and perilune and around Saturn, apochron and perichron. How flat the ellipse looks is called its eccentricity and is described in more detail in Chapter 1 – "Introduction".

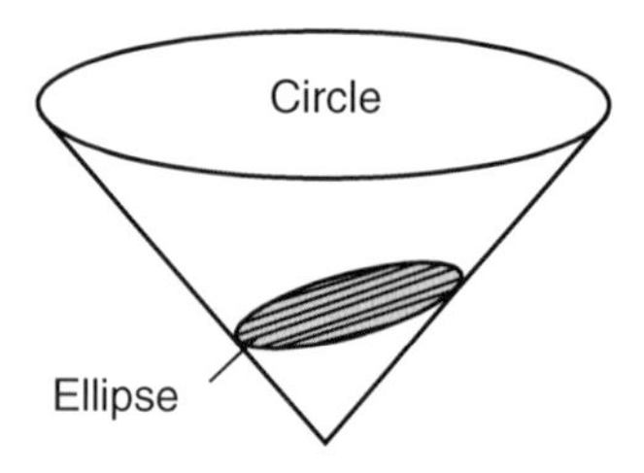

FIGURE 4.3 Ellipse Shape.

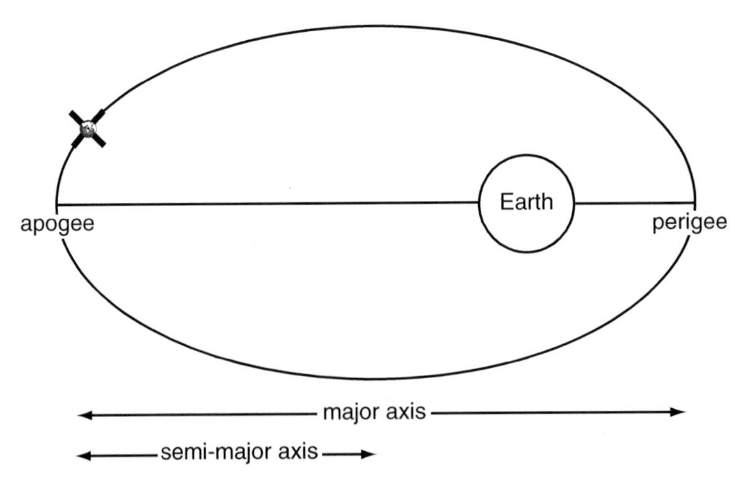

FIGURE 4.4 Properties of an Ellipse.

In an elliptical orbit, a satellite travels faster when it is closer to the object it is orbiting, and slower when it is further away. Johannes Kepler discovered this change in speed in 1609, by using data about the planet Mars' motion about the Sun, provided by the Danish astronomer Tycho Brahe. It became the second of his three laws that are now known as Kepler's Laws of Planetary Motion.

The first law states that the orbit of each planet is an ellipse, with the Sun at a focus. This also applies to all objects orbiting a body. For example, the centre of the Earth is always at a focus of an elliptical orbit of an Earth-orbiting satellite.

Kepler's second law describes that a planet travels faster when it is nearer the Sun than when it is further away, although the wording of the law make it sounds quite complicated. The law is stated as "The line joining the planet to the Sun sweeps out equal areas in equal times." This can be seen in Figure 4.5. Each of the shaded segments is equal in area to each of the un-shaded areas and the time for a planet to move from one area to the next is the same, no matter where on its orbit it is. Therefore, when it is nearer the Sun, a planet travels faster than when it is further away. Again, this not only applies to planets, but to all objects orbiting another body.

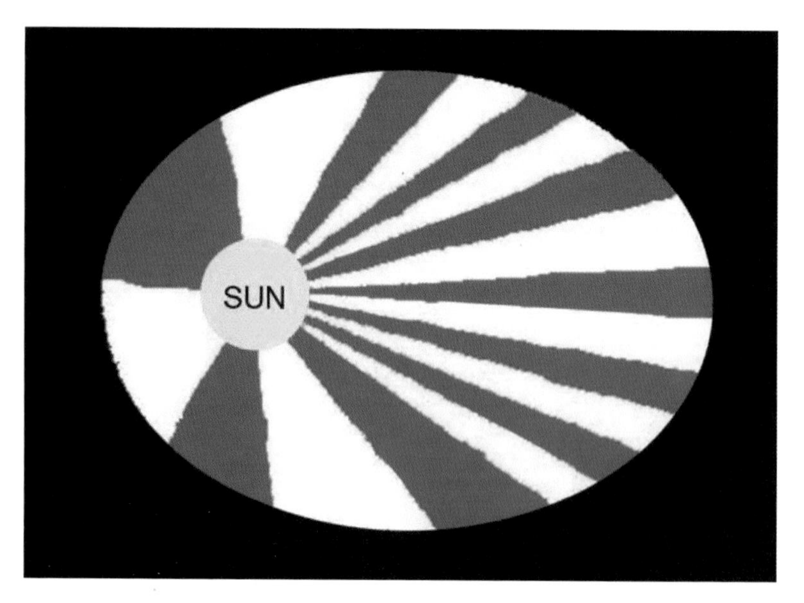

FIGURE 4.5 Diagram Illustrating Kepler's Second Law.

The third law sounds even more complicated, but just describes the fact that the length of a planet's year, or orbital period, is related to the average distance it is from the Sun. The relationship is "The square of the period of a planet is proportional to the cube of its mean distance from the Sun".

Each orbit can be described by six orbital elements or orbital parameters. These define the size, shape and orientation of an orbit and are described in Appendix A – "Orbital Elements".

Types of Orbit

Different types of orbit are used for different purposes, such as weather mapping, television broadcasts or manned space flight. All orbits around the Earth must circle the centre of the Earth and therefore they either follow the equator or cross it. The inclination of an orbit is the angle it makes when it crosses from the southern to the northern hemisphere. With improved technology and more advanced propulsion systems, it should eventually be possible to circle the globe above any line of latitude, although the amount of energy required may make it prohibitively expensive. If a spacecraft moves from west to east, it is said to be in a prograde orbit. This is the usual direction of rotation in our solar system. If it moves from east to west it is termed retrograde. Halley's comet has a retrograde orbit and Venus has a retrograde spin.

Low Earth Orbit (LEO)

The exact position of a low Earth orbit has not been clearly defined, but it is usually considered to be between 100 and 1,000 kilometres above the surface of the Earth. This is the cheapest and easiest orbit to put a spacecraft into. It requires less energy to get a spacecraft into a LEO than a higher altitude orbit, in a similar same way that it takes less energy to throw a ball to a height of 20 metres than it does to 30 metres. The position of the LEO is between the Earth's atmosphere and the inner Van Allen radiation belt. The radiation belts are rings of energized charged particles that are trapped by the Earth's magnetic field, described in more detail in Chapter 1 – "Introduction". The International Space Station, the *Hubble Space Telescope* and the *Iridium* constellation of satellites are all in LEO.

Geosynchronous Earth Orbit (GEO)

If a satellite is in a geosynchronous Earth orbit it travels once around the Earth in the same time that it takes for the Earth to rotate once, called a sidereal day. Satellites in geosynchronous orbits include weather satellites, pictures from which are shown on the news each night, and relay satellites, which NASA uses to relay communications and data between spacecraft, such as the International Space Station or the *Hubble Space Telescope* and control centres on Earth. If a spacecraft is in a GEO and the orbit is circular and directly above the equator, as opposed to an inclined orbit, it is called a geostationary orbit (GSO). A satellite in GSO will appear to hover over one point on the Earth all the time. This means that a receiving dish on the Earth pointing at that satellite will not need to move or track the satellite across the sky. This is the type of orbit many satellite TV channels are broadcast from, and why home receiving dishes in the northern hemisphere face south and do not have to move. A group or constellation of geostationary satellites can see the Earth's entire surface between about 70° south and 70° north.

All geostationary orbits must be geosynchronous, however, not all geosynchronous orbits are geostationary. If a geosynchronous satellite is not positioned over the equator or it does not have a circular orbit, it will not appear stationary, but will appear to move across the sky. When viewed from the surface of the Earth, a satellite in a circular orbit inclined at an angle to the equator will appear to travel north and south, in a figure of eight pattern. A satellite in an elliptical orbit will trace a slanted figure of eight pattern.

To maintain a geostationary orbit, a satellite must be about 35,880 kilometres above mean sea level. If the satellite were any higher, it would circle the Earth slower than the Earth takes to rotate, any lower and it would orbit quicker than the Earth's rotation. This can be illustrated by the Moon and the International Space Station, both of which have almost circular orbits. The Moon is about 385,000 kilometres away from the Earth and takes 27.3 days to complete its orbit. This is called a sidereal month. The International Space Station, however, is only about 390 kilometres above the Earth and takes about 90 minutes to complete an orbit.

The idea of the geostationary orbit has been discussed since the early part of the 20th century. Konstantin Tsiolkovsky, Hermann Oberth and the Slovenian rocket engineer Herman Potocnik, who was also known as Herman Noordung, all wrote about them. However, the English author and inventor, Sir Arthur C. Clarke, is credited with the first description of the orbit for use as a global communication system. In October 1945 he published an article titled *Extra-Terrestrial Relays* He suggested that communications around the world would be possible via a network of three geostationary satellites spaced at equal intervals around the Earth's equator. This idea was proved in 1964 when NASA's *Syncom 3* became the first geostationary satellite, and broadcast live pictures of the Olympic games in Tokyo, Japan. The geostationary orbit is sometimes referred to as the Clarke Orbit or Clarke Belt.

As the geostationary orbit is only at one altitude above the Earth, the number of satellites that can occupy geostationary positions is limited. It is further limited by the possibility of interference between the different satellite communication channels used to provide data between the Earth and the satellite. Another disadvantage of geostationary orbits is that there is a long distance between the satellite and the ground. This either means that, compared to satellites in lower orbits, more power or larger antennas are required for communication and it also takes a long time for a signal transmitted from a satellite in GEO to reach the ground. This is why geostationary satellites are no longer used for most two way telephone calls. The main benefits of a geostationary orbit are that it remains stationary relative to the Earth's surface and it can be seen from about a third of the Earth's surface. It is therefore ideal for communications and television broadcasts, as it is not necessary to track the satellite and move the antenna. This makes the geostationary orbit one of the most popular orbits and every year a lot of money is spent getting satellites there.

Polar Orbit

A polar orbit is where the satellite passes above or close to both the north and south poles of the Earth with each pass. As the Earth rotates below the satellite, the satellite will pass over a different region of the Earth with each cycle. Polar orbits are used to map the Earth and are also used by some weather satellites. These orbits are mainly at altitudes of about 1,000 kilometres, although some are as low as 200 kilometres, and therefore the resolution of the images sent back is much higher than from geostationary satellites. These orbits are usually circular and therefore they keep almost the same height above the ground all the time. The amount or swath of the Earth seen by the satellite during each pass is dependant on the instruments on board and the altitude of the satellite. Most meteorological satellites can cover the whole globe in one day as they have a swath of about 3,300 kilometres.

Sun Synchronous Orbit

A polar orbit that passes over the same part of the Earth at about the same local time each day is called Sun synchronous. This is useful in obtaining comparable data over a specific area. For example, air pollution over London, UK, at 9 a.m. will probably be very different to that at 9 p.m., due to variations in traffic movement at these times. To discover if the air pollution is improving or getting worse it is necessary to compare the quality at the same time each day. This could be achieved with a geosynchronous satellite, but most satellites in a Sun synchronous orbit are only about 600–800 kilometres above the Earth's surface and take about 95–100 minutes to circle the Earth.

Medium and High Earth Orbits

At around an altitude of 1,000 kilometres up to a geosynchronous orbit, is the Medium Earth Orbit. This is particularly useful for constellations of satellites used for telecommunications.

An Earth orbit higher than a geosynchronous orbit is sometimes referred to as a High Earth Orbit.

Highly Elliptical or Molniya Orbit

This type of orbit has an inclination of between 50° and 70°. It takes about 12 hours to complete one revolution. As a satellite travels faster at perigee and slows down at apogee, a satellite on this orbit swings by the Earth quickly, but then takes a long time over the higher altitude part of its orbit. The satellite is inserted into this orbit so that it will spend the greatest amount of time over a specific area of the Earth. For example, if communication is required in the arctic regions, a highly elliptical orbit communications satellite is launched whose orbit is configured so that is spends the bulk of its time above these latitudes. This type of orbit is also referred to as a Molniya orbit after the first Soviet communications satellites to use it.

Parking and Graveyard Orbits

A parking orbit is a temporary orbit where a spacecraft may wait for the correct timing until other celestial bodies or satellites are in the correct alignment for rendezvous or interception missions. It can also be used to wait for the delivery of components, spacecraft or the rectification of a fault.

A graveyard orbit is where a spacecraft is intentionally left at the end of its working life so that it clears the way for other operational satellites and reduces the risk of collisions and the generation of more space debris. For satellites in geostationary orbits, the graveyard orbit is a few hundred kilometres above the operational orbit.

Lagrangian Points

Kepler saw in 1619 that the closer a satellite was to the body it was orbiting the faster it travelled. Therefore, any spacecraft orbiting the Sun with a smaller orbit than the Earth not only has a shorter distance to travel around its orbit, but it also travels faster than the Earth. Such a spacecraft will not stay close to the Earth for very long, which can make communication between the spacecraft and Earth difficult. Similarly, if the orbit is larger than the Earth's, the spacecraft will begin to lag behind.

However, it is not just the Sun's gravity that affects a satellite in our solar system. All the planets and their moons also exert a gravitational influence. If the object is far enough away, the influence is minimal, however the interaction of two different celestial bodies on objects can produce some useful results.

There are five positions in space where the gravity of the Earth and the Sun interact to cause small objects placed at those points to stay put, relative to the Earth and the Sun. These positions are called the Lagrangian Points and are shown in Figure 4.6. They are named after the Italian-French mathematician Joseph Louis Lagrange who first calculated them in the late 18th century. They are sometimes also known as libration points. Lagrangian points occur between any two large bodies, and so there are Lagrangian points for the Earth and Moon and also for other planets.

Lagrangian Point 1 (L1)

If a spacecraft is between the Sun and the Earth, some of the force of the Sun's gravity is cancelled by the pull of the Earth's gravity

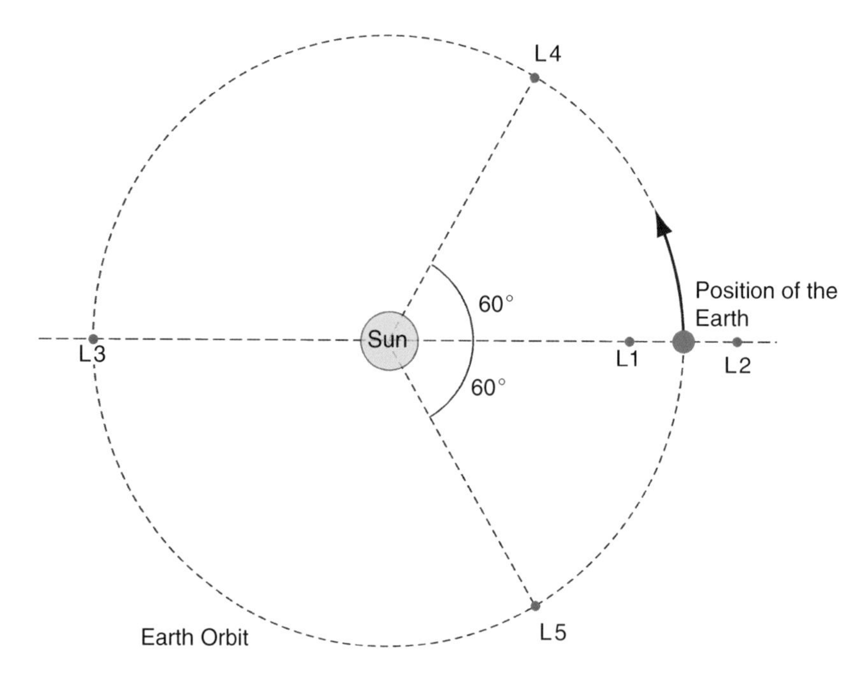

FIGURE 4.6 Lagrangian Points of the Sun and Earth.

in the opposite direction. It is the gravitational force that determines the speed an object requires to maintain a certain orbit, so when the effective pull towards the Sun is less, the spacecraft will need to travel slower than an object at the same distance from the Sun, but with no Earth nearby.

The reduction in the effective pull of gravity from the Sun will mean an object between the Earth and the Sun travels at a slower speed than it would otherwise, and at a certain point, the spacecraft will take exactly one year to orbit the Sun and it will therefore keep in the same relative position to the Sun and the Earth for the whole orbit. This is Lagrangian Point 1 and is about four times further away from the Earth than the Moon is. The *Solar and Heliospheric Observatory (SOHO)*, which is used to observe Sun's effects on the Earth, is located at L1.

Lagrangian Point 2 (L2)

At almost the same distance away from the Earth as L1, but further away from the Sun, is L2. This is where the effect of the Earth's gravity is added to the effect of the Sun's gravity and therefore means the spacecraft travels faster than objects in the same orbit but without the Earth's influence. L2 is where the ESA missions *Herschel, Planck, Eddington, Gaia,* the *James Webb Space Telescope* and *Darwin* are planned to reside.

Lagrangian Point 3 (L3)

This is just outside the orbit of the Earth, but on the opposite side of the Sun, and therefore twice as far away as the Sun. In comparison, L1 and L2 are a hundredth of the distance of the Sun away from the Earth. As the Sun is directly between the Earth and L3, no object that is orbiting there can be seen from the Earth. Because of this and the associated communication problems, no missions are currently planned for L3.

The L1, L2 and L3 Lagrangian points are unstable. It is like placing a ball at the top of a hill. A small push and it will roll away. Satellites at L1, L2 or L3 will drift away from the point, due to the gravitational influences of other planets and the Moon. They must therefore be capable of correcting their position.

Lagrangian Points 4 and 5 (L4 and L5)

These points are on the Earth's orbit, but they are 60° or about 148 million kilometres in front of and behind the planet as it travels around the Sun. Unlike Lagrangian points L1, L2 and L3, objects orbiting at L4 or L5 are stable and resistant to gravitational perturbations. These two points act like a bowl. If a ball is placed in a bowl, even if it is disturbed, it comes back to the centre of the bowl. In a similar way, if an object at L4 or L5 begins to drift away, it will always drift back to the same point. Because of this stability, objects tend to accumulate in these points and astronomers have discovered large asteroids at the L4 and L5 points of the Jupiter and Sun system. There are no asteroids at the Earth and Sun L4 and L5 points, however it is thought that dust and other debris may have accumulated there. It has been suggested that future space colonies could be located at L4 or L5, as keeping such a large construction in one place in any other orbit would require a vast amount of energy.

Getting into Orbit

The Earth's rotation can be used to help launch spacecraft, as explained in Chapter 3 – "Space Missions". To take the most advantage for the Earth's rotation, a satellite should be launched due east from the equator. This is one of the reasons why the French Guiana Space Centre at Kourou, at 5.2° North is well placed. An eastward launch gives a satellite the initial inclination of its orbit, which is almost equal to the latitude from where it was launched. So from Kourou, this would be about a 5° inclination. From the Baikonur Cosmodrome in Kazakhstan it would give about a 46° inclination. If a launch is not due east, then the inclination will be greater than the launch site latitude. To change the inclination once in orbit requires a lot of fuel. A polar orbit can be achieved from any latitude, by launching towards the north or south.

The point at which a satellite enters orbit is called the injection point and the speed at this point is called the injection velocity. If the injection velocity is insufficient for a circular orbit to be achieved, the resulting orbit will be parabolic. If the perigee is within the Earth's atmosphere, or below the Earth's surface, the satellite will fall back to Earth. The Russians called this injection velocity the first cosmic velocity. If the injection velocity is greater than that required for a circular orbit, an elliptical orbit will occur, but it will not fall back to Earth. If the injection velocity is greater than the escape velocity of the Earth, which the Russians called the second cosmic velocity, the satellite will leave the Earth's orbit and start circling the Sun. The Russians call the Sun's escape velocity the third cosmic velocity.

Changing Orbits

The orbit of most spacecraft will have to be changed at sometime in its life. This could be from a parking orbit to the final orbit or trajectory, to avoid space debris or micrometeors, as protection from solar storms or to correct an orbit that has changed due to perturbations. Also, at the end of the spacecraft's life, it may be transferred to a graveyard orbit.

Perturbations

The main influence on an orbit is usually the gravity of the body being orbited. However, other, usually much smaller, influences can change the orbit. These changes are called perturbations and can be caused by such things as atmospheric drag, the influence of gravitational forces of other celestial bodies, such as the Moon or planets, and also by solar activity. Also, the object being orbited may not be completely spherical or the mass may not be evenly distributed throughout the body. This may also cause perturbations, known as "perturbations due to non-sphericity". The orbits of satellites can also be perturbed by solar radiation pressure and infrared radiation. Large, light satellites, such as *Echo 1*, which was an aluminium coated balloon with a diameter of about 30 metres, suffer most from solar radiation pressure. Satellites that are close to the Earth can also be disturbed slightly by the tides of the oceans and it is thought that even the reflection of sunlight from the atmosphere can cause small perturbations. All these perturbations cause the orbit to no longer be truly elliptical, and when precise determination of an orbit is required, these perturbations must be taken into account. At low altitudes, atmospheric drag slows a satellite down. Even at altitudes of tens of thousands of kilometres, the Earth's atmosphere is still in evidence albeit in a very rarefied form, and this also slows satellites. For an elliptical orbit, at perigee the satellite is at its nearest to the Earth, and therefore the atmospheric drag is most noticeable. Over time, the drag will cause an elliptical orbit to become circular. Drag on satellites in circular orbits will cause them to slow down further, which will reduce their altitude until eventually the satellite will either burn up in the atmosphere or reach the Earth.

Station-keeping

Changes in the orbit due to perturbations need to be corrected by manoeuvres called station-keeping. This is done periodically over the lifetime of the spacecraft. In a geostationary Earth orbit, the combined gravitational forces of the Sun and Moon cause the orbital inclination to increase by about 1° a year. This is usually countered by a manoeuvre about once every two weeks, to keep the satellite aligned with the equator. During the lifetime of the satellite, this uses about 95% of the total amount of fuel supplied for station-keeping. The equatorial bulge of the Earth makes a satellite drift longitudinally, and a manoeuvre once a week will keep it above the correct longitude. Many satellites now use ion propulsion drives for their station-keeping.

Changing the Altitude or Shape of an Orbit

The easiest type of orbit change is when the desired orbit is in the same plane or inclination as the original orbit, this is known as coplanar. It requires less fuel than any other orbital change. If a satellite is slowed at perigee, the altitude of the apogee will drop, without affecting the perigee altitude. This is what happens when drag on a satellite changes an elliptical orbit into a circular one. Similarly, if a satellite's speed is increased at perigee, the apogee altitude will increase, again without affecting the perigee altitude, as shown in Figure 4.7. The same applies at the apogee. A change in speed at the apogee will change the altitude at perigee, without affecting the apogee altitude. These changes of speed are usually performed by a thrust on the rocket from a propulsion system.

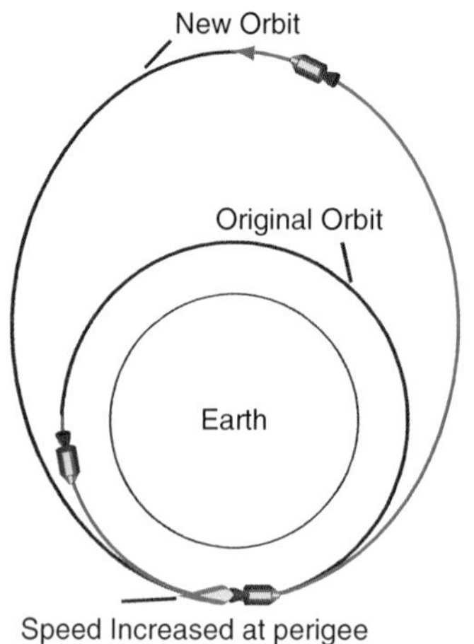

FIGURE 4.7 Orbital Change by an Increase in Speed at Perigee.

To increase the speed, the thrust received by the spacecraft must be in the direction of travel and therefore the propellant is ejected in the opposite direction. To decrease speed the thrust must in the opposite direction to that of travel and the propellant expelled in the direction of travel.

To change from one circular orbit to one with a different altitude but with the same inclination, the Hohmann transfer is usually used, shown in Figure 4.8. This is the most fuel-efficient method. To get to a higher circular orbit, a thrust is applied at perigee, in the direction of travel. This speeds the satellite up and causes the orbit to change to an ellipse. At the apogee of the new orbit, another thrust is applied, again in the direction of travel, this also speeds the satellite up. This time however, it increases the altitude of the perigee. With the correct impulse, or the correct amount of thrust for the correct amount of time, the new orbit will be circular and have the same inclination as the original orbit, just at a higher altitude. Although this is the most fuel-efficient way to perform the manoeuvre, it takes the longest time. The German civil engineer Walter Hohmann first described this manoeuvre in

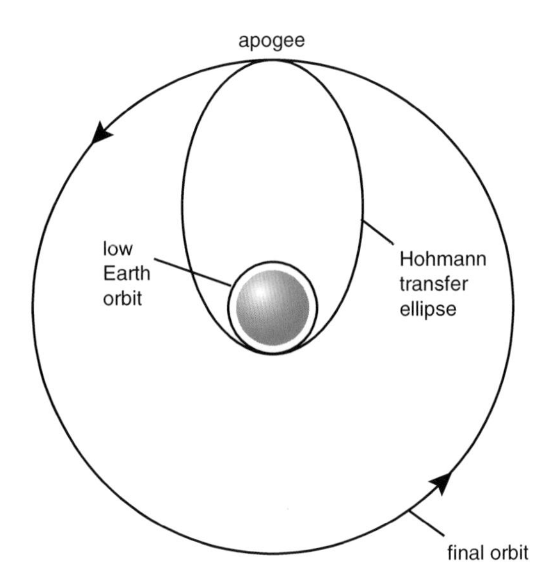

FIGURE 4.8 Hohmann Transfer.

1925. As the engines must fire twice, it is known as a two-impulse transfer. This type of transfer is often used to change the orbit of a satellite from a low Earth orbit into a geosynchronous orbit, called a geosynchronous transfer orbit.

To change the altitude of an orbit more quickly, other two-impulse transfers can be used, although they use more fuel. If the initial impulse is greater than required the transfer orbit will intersect the outer orbit more quickly, but then the second impulse must counter the extra outward movement and be fired inwards to establish a circular orbit, as is shown in Figure 4.9.

Another method is to angle the initial impulse outwards of the orbit, and when the new ellipse intersects the desired orbit, to fire back in, as shown in Figure 4.10. These can also be used for general transfers between elliptical orbits.

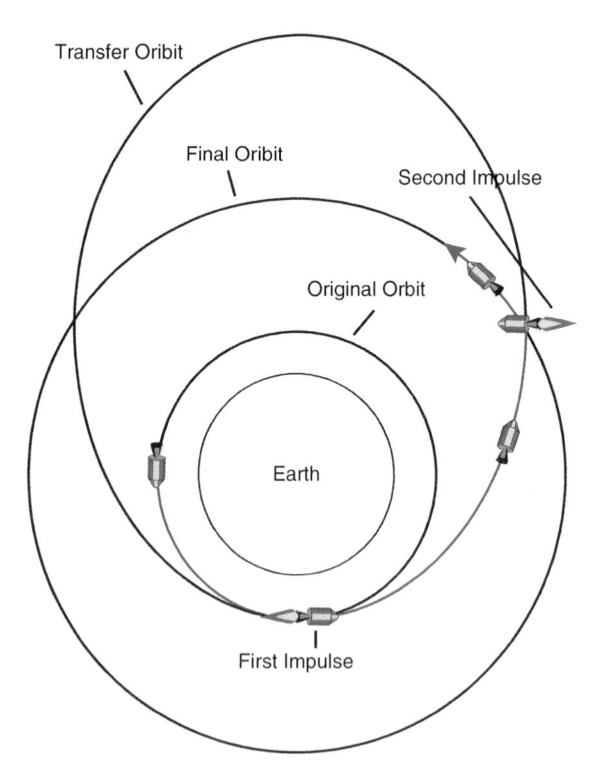

Figure 4.9 Two Impulse Transfer.

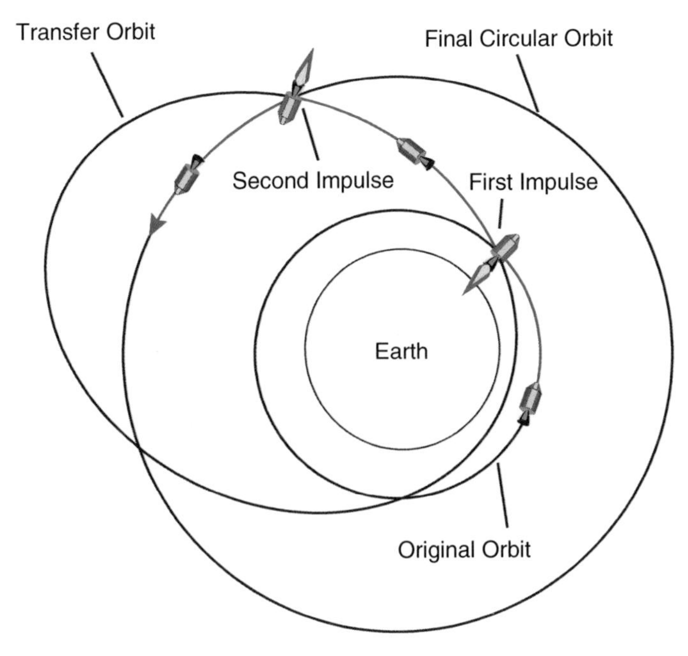

FIGURE 4.10 General Transfer.

Low Thrust Trajectory

Some spacecraft do not rely on just one or two impulses, but instead are propelled either continuously or in pulses, usually by an ion drive. Every time the ion drive fires, the spacecraft is nudged into a slightly different orbit, which, over time, changes the trajectory significantly. This method can result in either a quicker journey time, or will require less fuel to reach certain celestial objects than a conventional rocket would use. At first, the trajectory of an ion-propelled craft looks like a Hohmann transfer orbit, but after long periods of continuous operation the trajectory will no longer be a purely ballistic arc. These types of trajectory are quite complex to model as the force is continually changing. ESA's *SMART 1* spacecraft used an ion propulsion drive when it left the Earth and also when it approached the Moon, and its path to the Moon can be seen in Figure 4.11.

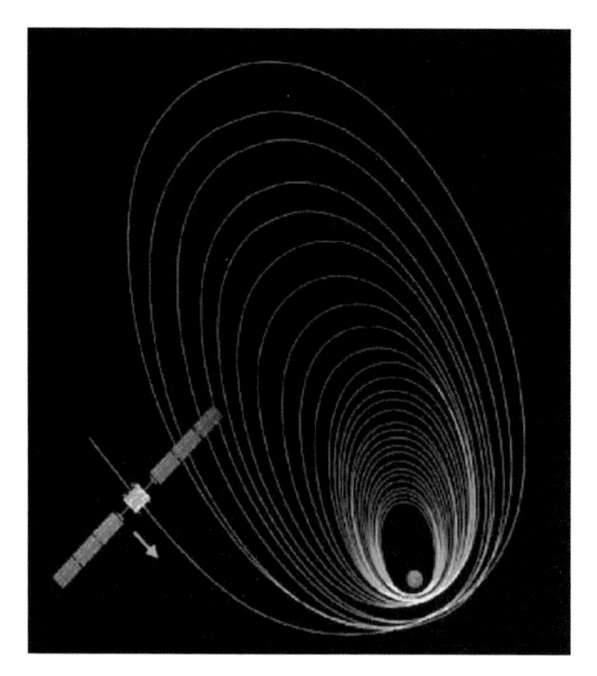

FIGURE 4.11 *SMART 1's* Trajectory Near the Moon.
Image courtesy ESA

Orbit Inclination Changes or Out-of-plane Orbit Changes

If only the plane or inclination of an orbit is changed, with the size and shape remaining constant, it is called a simple plane change. This type of change is used to change an inclined orbit to an equatorial orbit, for example. In order to change the inclination a thrust must be applied at the point where the original orbit crosses the plane of the required orbit. Therefore if an inclination of zero is required for an equatorial orbit, the thrust must be applied over the equator. This can be either where the path changes from the northern to southern hemisphere or from the southern to northern hemisphere, known as the descending and ascending nodes. The impulse required to produce a change of inclination is very large, but the speed of the satellite when it is in the final 0° orbit will remain the same as it was in the inclined orbit. As the impulse required depends on the speed of the satellite, plane changes for objects in a low Earth orbit need much more fuel than those in a higher Earth orbit. In some low Earth orbit cases, it is worth transferring to a higher orbit by using a Hohmann transfer, performing the plane change and then reducing the size of the orbit back again by another Hohmann transfer.

Rendezvous and Intercept

To get a spacecraft to the correct place at the correct time, for example, to allow the Space Shuttle to dock with the International Space Station, or for a probe to land on Mars, requires analysis to determine the optimum timing and sequence of manoeuvres. To minimize the cost of the mission, which can be measured in terms of change in velocity or characteristic velocity, computer analysis is used to provide the optimum launch time, the manoeuvres and the time-of-flight, which is done at the mission planning stage.

There are four stages to a rendezvous. The first stage is the launch of the spacecraft into space and its injection into the rendezvous trajectory that will carry it into the vicinity of the target. The second stage is simply its journey along the trajectory, usually called the coasting stage. The third stage begins when it nears the target, when the precision manoeuvres bring the spacecraft almost into contact with the target. Finally, the fourth stage consists of either docking the spacecraft with the satellite and linking the two space vehicles together, or landing the spacecraft on the other body, such as on Mars.

Rendezvous in space is not a simple matter of just speeding up to catch up with something. Figure 4.12 shows two spacecraft,

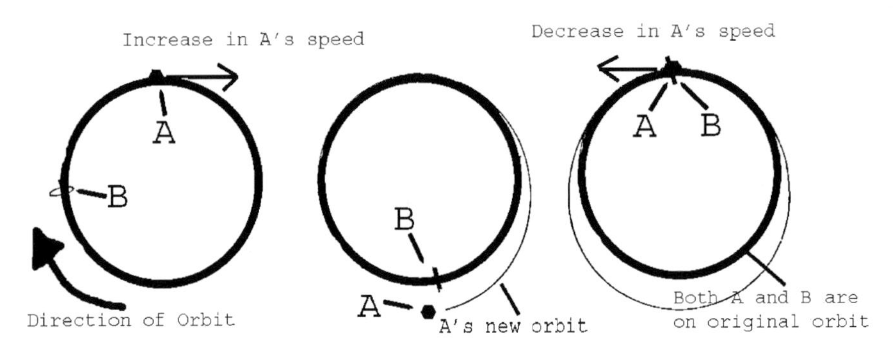

FIGURE 4.12 Rendezvous Manoeuvre.

A and B on the same orbit, but B lags behind A by about a quarter of an orbit. For them to be able to dock together, it would seem logical for A to slow down, so B can catch up. However, to use the minimum amount of energy, A actually speeds up, to put it into an elliptical orbit. It then takes longer to orbit than B, so after one or more revolutions, they coincide. At this point, spacecraft A is slowed down, so that it regains its original circular orbit and the two spacecraft can then dock.

Interplanetary and Interstellar Trajectories

Most space missions are designed to use the minimum amount of propellant so the costs are not prohibitive and the mass of the craft and propellant does not exceed the capability of the launch vehicle. The route to the destination is the main factor in how much propellant is needed, and so those trajectories that need the least fuel are of the most interest.

To get from one planet to another, called an interplanetary trajectory, involves escaping the original planet's gravity. The spacecraft will then be in an orbit around the Sun, in interplanetary space. Everything in space is moving around in its own orbit, and everything is moving in relation to everything else. The distance between any two bodies changes with time, for example, the closest Mars ever comes to the Earth is about 55 million kilometres but it can be as far away as 400 million kilometres.

For a spacecraft to travel from the Earth to Mars it must trace a whole or partial orbit around the Sun combined with either Hohmann or other orbital transfers, as illustrated in Figure 4.13. It must be remembered that if Mars were in one position when a spacecraft was launched, by the time of rendezvous Mars would have moved further around its orbit. To reach Mars is like throwing a dart at a moving target, whilst standing on a moving platform. You have to aim at not where it is now, but where it will be when you get there. Because of the different orbits of the Earth and Mars, the opportunity to launch a spacecraft to Mars on a transfer orbit, such as a Hohmann transfer, occurs about every 25 months. To get to Venus there is a launch opportunity about every 19 months.

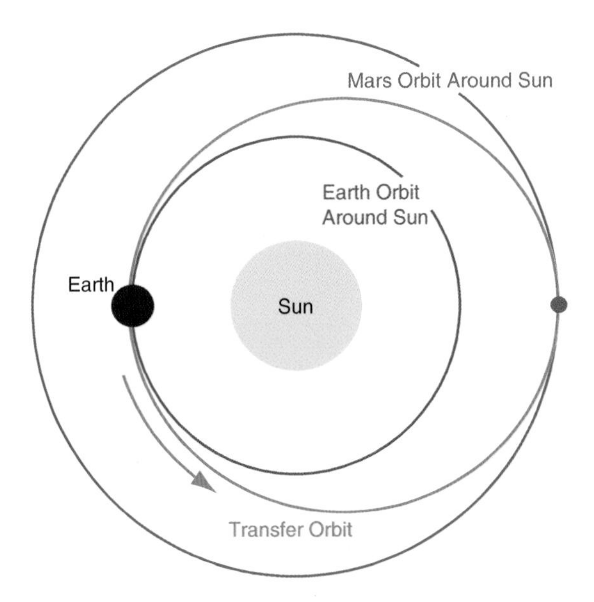

FIGURE 4.13 Trajectory to Mars from the Earth.

When the spacecraft is in the vicinity of its destination or target, it must then be slowed down so that the target's gravitational field can capture it. The smaller the target, the weaker its gravitational field and the more the spacecraft must slow down.

Gravity Assist, Fly-by or Sling Shot

After the propulsion system has stopped, a spacecraft's trajectory is determined by gravity. There is a maximum speed that the spacecraft can reach by using propellants, and this is limited by the amount of fuel the rocket can launch. With present launch technology, not enough fuel can be lifted from the Earth's surface to reach the Sun's escape velocity and so the distance that a spacecraft can travel away from the Sun, relying solely on propellants, before it starts to head back on an elliptical orbit, is limited. Currently this limit is a bit beyond the orbit of Jupiter and therefore the outer solar system is inaccessible by propulsion alone. As propellant is also used to slow a spacecraft and therefore reduce its distance from the Sun, the planet Mercury, which has a smaller orbit than the Earth, is also impossible to reach by just using current propulsion technology.

Therefore some interplanetary or interstellar journeys do not rely just on propellants to move them nearer to their destination. If a spacecraft passes close to another planet, it can use the gravitational pull of the planet to increase its speed or change its direction of travel. This is known as gravity assist, fly-by or a sling shot manoeuvre. When a spacecraft leaves the Earth's orbit, it follows an orbit around the Sun. If it then approaches another planet, the gravity of that planet pulls the spacecraft towards it and alters the spacecraft's speed and direction. The spacecraft also alters the speed of the planet, but this change is imperceptibly small. The direction the spacecraft approaches the planet, and whether it passes in front of or behind it, determines the amount the spacecraft speeds up or slows down. As long as the spacecraft is still travelling faster than the escape velocity of the planet when it leaves the planet it will follow a different orbit around the Sun, unless it has sped up fast enough to leave the Sun's influence, in which case it will head off into interstellar space.

As the spacecraft leaves the planet, if it had sped up, it begins to slow again. If someone were watching from the planet, it would appear that the spacecraft sped up and slowed down the same amount, similar to a cyclist speeding up coming down a hill and slowing down again as they went up the other side. As the planet is moving around the Sun, it has angular momentum. Momentum is a measure of the amount of motion an object has. Angular momentum is how much rotation an object has travelling around a point, whereas linear momentum, often just known as momentum, is how much motion an object has travelling in a straight line. As the spacecraft approaches the planet, it gains some of the planet's angular momentum. The total angular momentum of the spacecraft and the planet always remains the same, known as conservation of angular momentum. As the spacecraft leaves the planet, it keeps the angular momentum it has acquired, and it speeds up and the planet slows down, but as the planet is so massive, this change in speed is imperceptible. Viewed from the Sun, the speed of the spacecraft before it approached the planet is slower than after it leaves. A similar effect is seen if a ping-pong ball is thrown towards a ceiling fan. As the ball bounces off of the blade, it shoots out faster than it was thrown in. The fan slows slightly, but soon regains its speed from its motor. NASA's *Voyager 1* and *2* spacecraft used gravity assist to change their speed and direc-

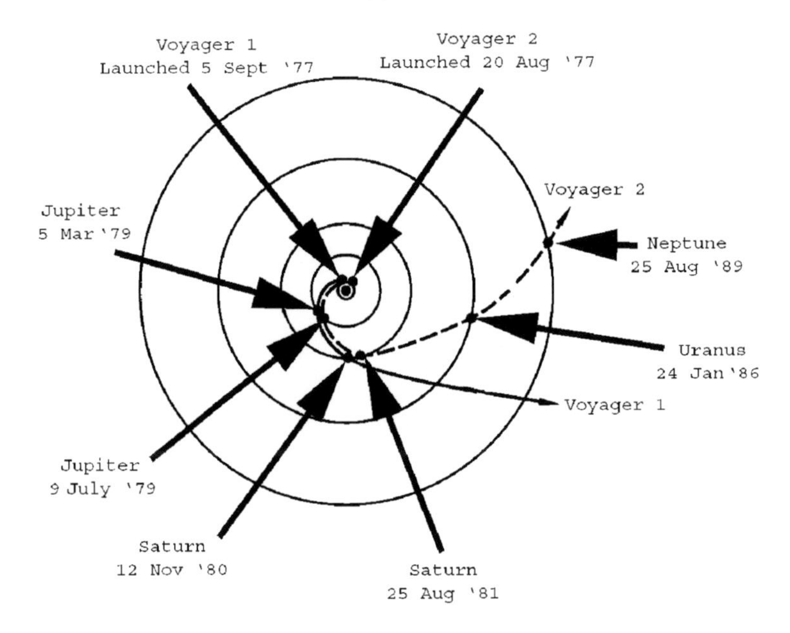

FIGURE 4.14 *Voyager 1* and *2* Trajectories.
Image courtesy NASA

tion, as can be seen in Figure 4.14. The *Cassini Huygens* mission to Saturn used two gravity assists from Venus, followed by one from the Earth and one from Jupiter, before it reached Saturn.

The orbit of the spacecraft can be changed dramatically using gravity assist and it can even shoot out of the plane of the ecliptic. The spacecraft *Ulysses*, which is used to explore the region of space above the Sun's poles, first travelled out towards Jupiter and was swung up by Jupiter's gravity out of the ecliptic and into a solar polar orbit.

In the early 1960s, Michael Minovitch championed the idea of gravity assist while he was a UCLA graduate student. The first spacecraft to use a gravity assist was NASA's *Pioneer 10* in 1973. It approached Jupiter travelling at almost ten kilometres per second. After it passed Jupiter, it sped off into deep space at about 22 kilometres per second.

Gravity assists can also be used to slow down a spacecraft. If a spacecraft flies in front of the planet, it will donate some of its angular momentum to the planet and therefore slow down. The Italian astronomer Giuseppe "Bepi" Colombo had described this method of deceleration, and designed a "Mission Impossible" to Mercury, the innermost planet of our solar system. To reach Mercury, a spacecraft launched from Earth needed to lose more energy than a conventional rocket would allow, so in 1974, NASA's *Mariner 10* spacecraft flew past Venus first and lost some of its speed. It then entered its rendezvous orbit with Mercury.

Ground Tracks

The point on the Earth's surface directly below the satellite is known as the subsatellite point, and the path on the ground that it follows during its orbit is called the ground track. The subsatellite point is where an imaginary line taken from the centre of the satellite to the centre of the Earth hits the Earth's surface. If only the altitude of the satellite's orbit is changed, the ground track remains the same. The position of the ground track is useful for many reasons, including weather mapping and communications. If meteorologists are trying to monitor a developing hurricane, they need to know which satellite will pass over the area they are interested in and when. Similarly, ground stations will need to know which satellites are passing overhead so that they can communicate with them.

The ground track is usually plotted on a Mercator projection map of the Earth. This type of map flattens the Earth from a globe into a rectangle. The lines of longitude are from the top to the bottom of the rectangle, and the lines of latitude go from left to right. A circular orbit will produce an s-shaped curve, as seen in Figure 4.15, the maximum latitude north and south of the equator being the same as the inclination of the orbit. If the satellite's orbit is retrograde, that is, it goes the opposite way to that which the Earth is turning, the maximum latitude is 180° minus the inclination, due to the way the inclination is measured.

As the Earth rotates underneath the satellite, after one orbit the satellite will not be above the same position on the Earth. Where it is depends on the period of the orbit, or how long one orbit takes, and whether the satellite is travelling prograde, that is, the same

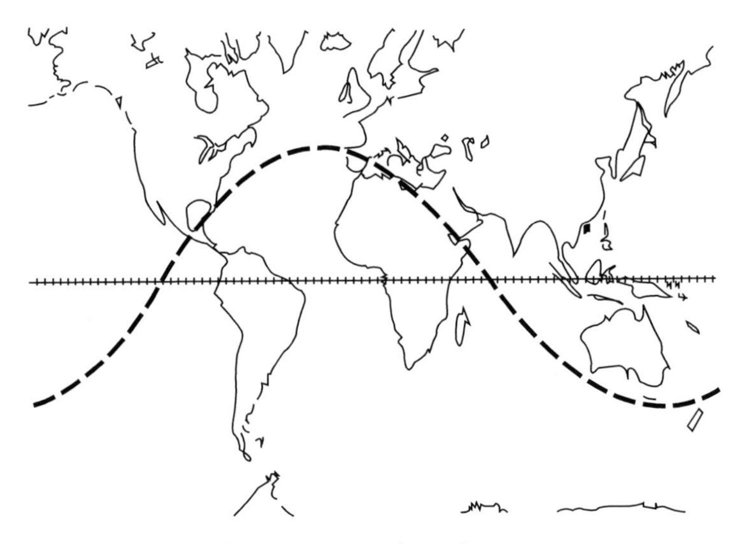

FIGURE 4.15 An S-Shaped Curve Ground Track.

way that the Earth is turning, or retrograde, the opposite way. The length of time for the Earth to rotate once is 23 hours and 56 minutes, known as a sidereal day. Therefore, in one hour the planet turns a tiny bit over 15°. If the time to circle the Earth for a prograde satellite in a circular orbit is two hours, the ground track will be just over 30° further west during the next orbit as can be seen in Figure 4.16. If the period of the orbit is a multiple of the sidereal day, then the satellite will eventually retrace exactly the same path as it did on its first orbit.

If a satellite is in a polar orbit, the ground track will appear as a slanted straight line, as can be seen in Figure 4.17. As the Earth rotates about 15° in one hour, so a satellite in a polar orbit with a two hour orbit, will have a ground track that will slant westwards by about 15° as the satellite travels from one pole to the other. Each orbit will therefore start 30° further west than the previous orbit.

Elliptical orbits, such as geostationary transfer orbits and the Molniya orbits used for communication over high latitudes, are not the characteristic "s" shapes of the circular orbits, although the maximum and minimum latitude is still the same as the inclination of the orbit. Their characteristic ground tracks can be seen in Figures. 4.18 and 4.19.

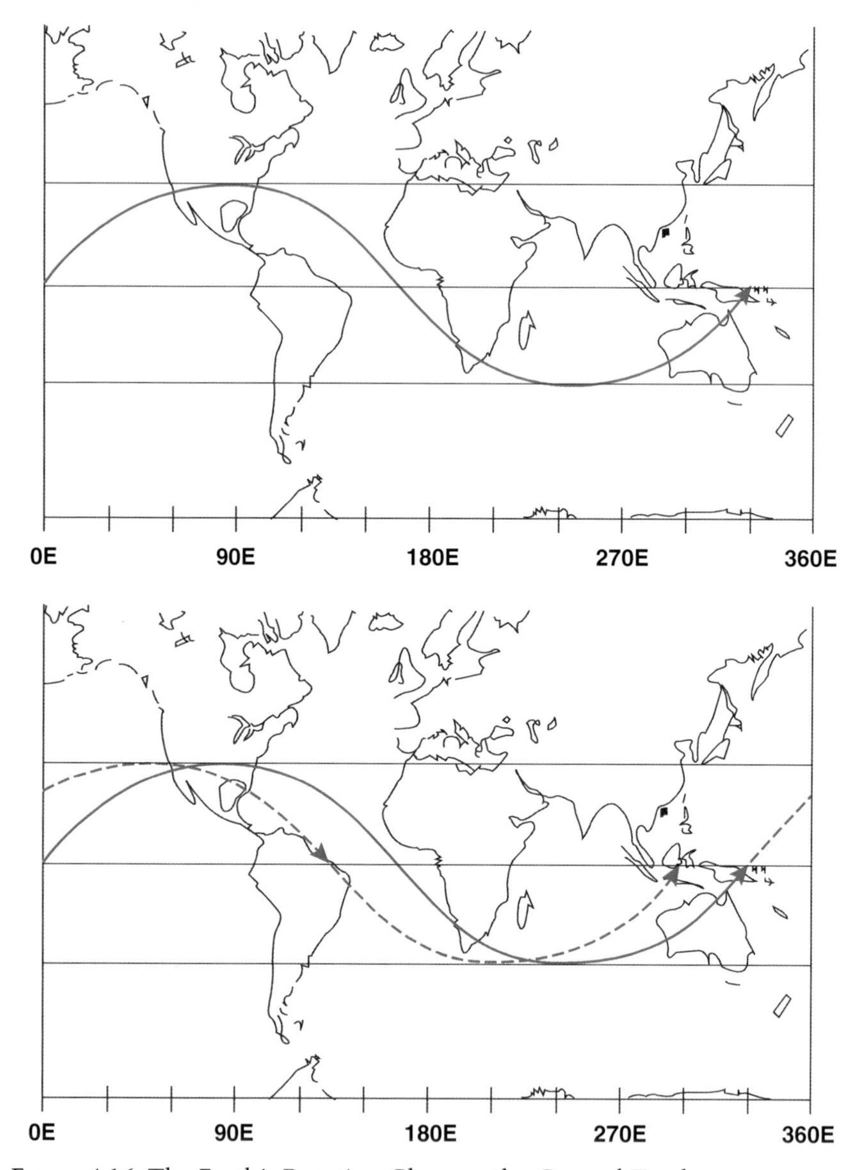

FIGURE 4.16 The Earth's Rotation Changes the Ground Track.

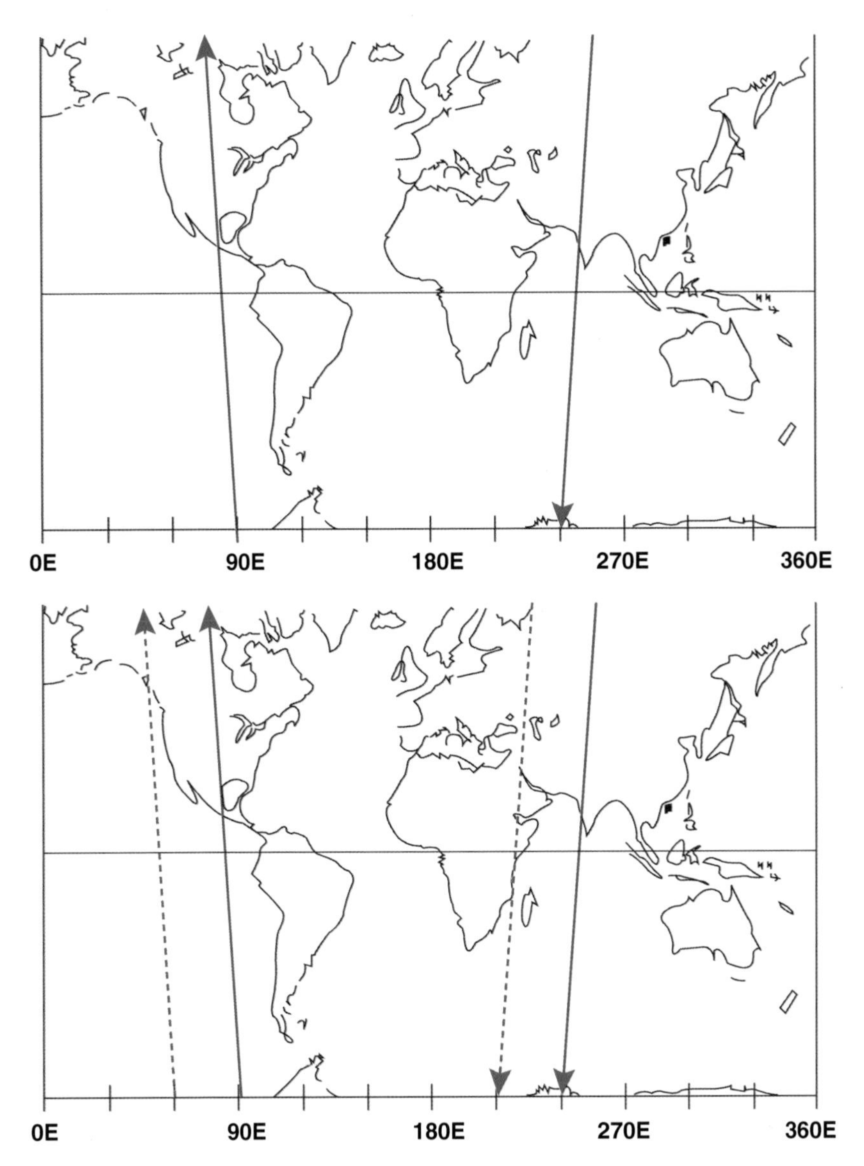

FIGURE 4.17 Polar Orbit Ground Track.

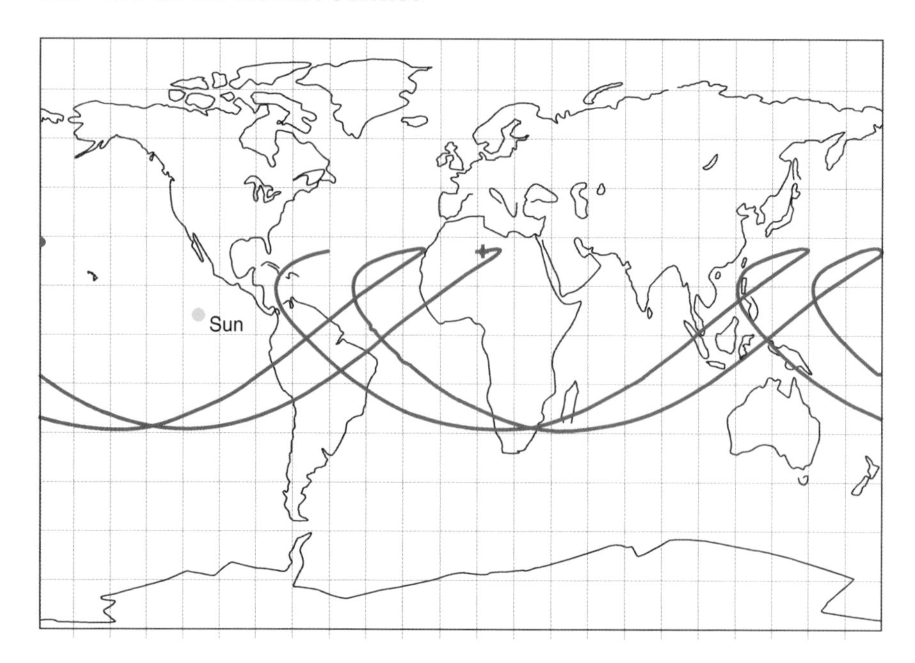

FIGURE 4.18 Elliptical Orbit Ground Track.

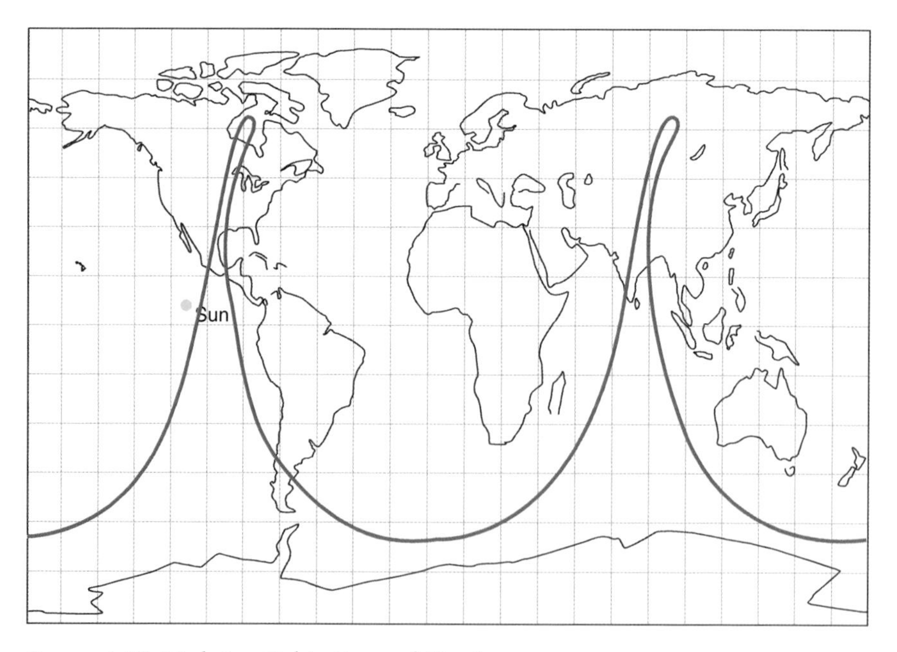

FIGURE 4.19 Molniya Orbit Ground Track.

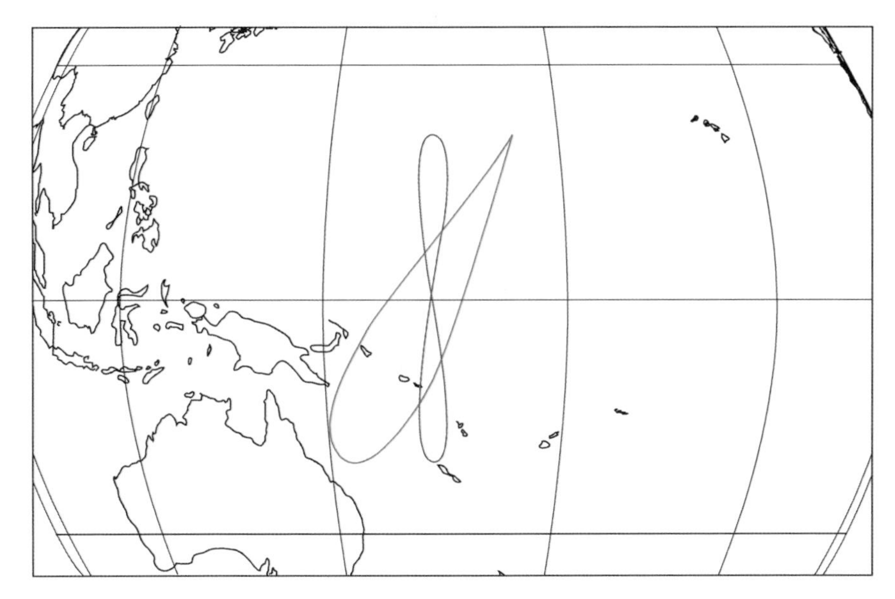

FIGURE 4.20 Geosynchronous Orbit Ground Track.

A geostationary orbit should have a ground track of just a single point, however, perturbations cause the orbit to change shape slightly and so the ground track is often a small figure of eight pattern. Geosynchronous orbits that are not geostationary trace a larger figure of eight pattern, as can be seen in Figure 4.20.

5. Propulsion Systems

Propulsion systems are required to do three things. First, they must provide lift for the launch vehicle and its payload, and take it from the launch pad up into space. They must then ensure the payload travels around the Earth, so that it can enter a low Earth orbit or a parking orbit, and not fall back down to Earth. The third task is to manoeuvre the payload into the desired height, position or trajectory.

The propulsion system on the launch vehicle is also often responsible for moving the payload straight into the parking orbit. Once in orbit, other propulsion systems are usually used to transfer the payload from the parking orbit into a higher orbit or into a trajectory for planetary encounters. As the movement of the spacecraft around its orbit already balances the force of gravity, and as it does not need to battle with air resistance, propulsion systems that produce a small force or thrust can be used for this third stage.

The amount of thrust produced by each type of propulsion system is mainly determined by how much and how fast the exhaust leaves the exit. A large thrust usually requires a lot of exhaust leaving very fast, a lower thrust means less or a slower exhaust. The specific impulse, usually abbreviated to Isp, is a measure of a propulsion system's power and efficiency at using the propellant. It can be thought of as similar to the litres per 100 kilometres rating of a car, but Isp has the units of seconds. The higher the value of Isp the better, as the more thrust that can be produced per kilogram of propellant the less fuel it needs to carry. The total impulse can be thought of as a measure of stamina or how much thrust can be produced and for how long. As the total operation time is dependent on the amount of propellant carried, total impulse is a measure of the complete vehicle, including the size of the fuel tanks, and not just the propulsion system.

There are various methods of getting a propellant to produce thrust. The main three are a chemical reaction used in chemical propulsion systems, a nuclear reaction for nuclear propulsion systems and an electric propulsion system, which uses electricity.

Chemical Rocket Propulsion

Chemical rocket propulsion is the only method we currently have to escape from the surface of the Earth. No other type of propulsion system can produce enough thrust to overcome the gravity on the Earth's surface. Launch vehicle propulsion systems must have a high thrust and therefore produce a lot of force, but they are usually only required to run or burn for a short time such as a few minutes.

Chemical rockets not only work in space, but they also work in the atmosphere. However, until the launch vehicle has risen above the atmosphere it must also push the surrounding air out of the way, which wastes some of the rocket's energy, and rockets are therefore less efficient in the atmosphere than in the vacuum of space.

Chemical propulsion systems use chemical propellants. Before they are burnt, these can be solid, liquid or gas or they can be used in combination with each other. Where two different types of propellant are used, the system is called a bi-propellant. For historical reasons, propulsion systems that use solid propellants are called motors, and those that use liquid propellants are called engines or thrusters. Different types of propellant are used for different purposes and the appropriate selection is important. Some propellants produce toxic fumes and others cannot be easily stored and are not suitable for long duration missions.

Most chemical propellants need a fuel and an oxidiser. The oxidiser contains all of the oxygen required to allow the chemicals to react or burn and therefore an external oxygen supply is not required. This is different to a bonfire on the Earth, which utilizes the oxygen in the air to burn. Some chemical propellants however, just require a catalyst before they decompose to produce heat and gas.

Solid Propellants or Rocket Motors

Solid propellants are by far the simplest type of rocket propulsion system. A firework rocket is very similar in design to a rocket motor. Figures 5.1 and 5.2 shows a section through a firework and

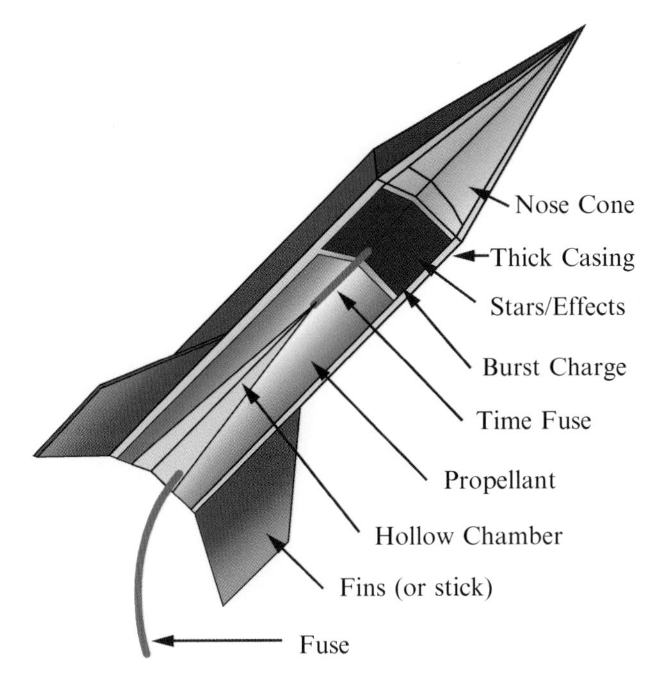

FIGURE 5.1 Section Through a Firework.
Image courtesy Colin Bradley http://www.pyrouniverse.com

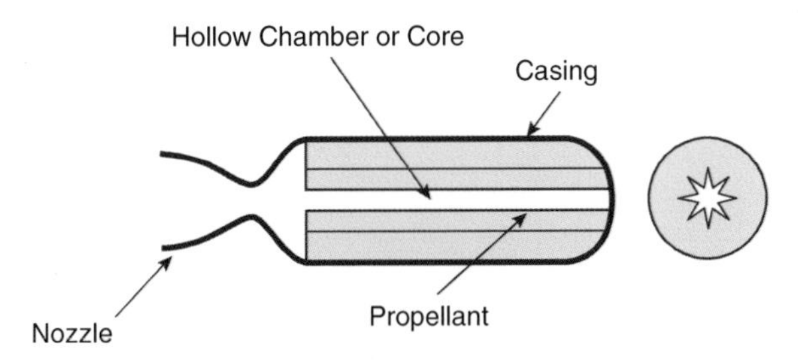

FIGURE 5.2 Section Through a Solid Rocket Motor.

a simplified solid rocket motor. If ignited, a firework would travel in space or even underwater, assuming the powder was kept dry.

The British *Skylark* research rocket, which flew from 1957 until its final launch in Sweden on May 2, 2005, used solid rocket motors. NASA's Space Shuttle also uses two solid rocket motors, or boosters, strapped on either side of the main engine. They are the largest solid propellant motors ever to fly and are over 45 metres long and three metres in diameter. Solid propellants can be stored for periods of between 5 and 25 years without deterioration and can be transported easily, but they cannot withstand damage or being dropped. This type of motor requires very little servicing but cannot be tested before the final ignition. Most solid propellant rockets require an external ignition system, similar to lighting the fuse on a firework.

Solid Rocket Motor Design

The solid propellant is a mixture of chemicals consisting of the fuel, oxidiser and a binding material. These are bonded together to form a compound called the grain, which is then moulded around a core. The core is then removed leaving a cavity in the grain, called the cavity core. The motor's performance is determined by the chemicals used in the grain and also by the interior shape of the cavity core and by the nozzle design. Typical chemicals used are powdered aluminium and ammonium perchlorate. The aluminium is the fuel and the ammonium perchlorate is the oxidiser. The chemicals do not ignite when mixed together, and a synthetic rubber-like material such as polyurethane or polybutadiene is used to bind them together into the required shape. Solid rocket motors can vary in thrust from about two newtons to over four million newtons. One newton, abbreviated to 1 N, will lift 100 grams, or a small apple, from the surface of the Earth. Simple solid rocket motors have no moving parts, however some of the more complicated designs incorporate moveable nozzles.

The ignition system is composed of an igniter propellant that is electrically lit. The igniter propellant can be made from the same chemicals as those in the main grain, or from different chemicals. The burning igniter propellant produces a lot of heat and gas very quickly, which rapidly fills the cavity in the grain and causes the grain to ignite.

The casing is designed to withstand the great pressures and temperatures produced within the motor. It is usually made of a metal such as steel, aluminium or titanium, or a composite fibre-reinforced plastic. The lighter the casing, the more efficient the motor, as the casing must also be lifted. To protect the casing from the heat inside, it is usually insulated, although the grain itself also insulates the casing during the initial stages of burning. The hot gases produced by the combustion of the propellant flow though the cavity and out through the nozzle. The nozzle is designed to expand and accelerate the hot gases to supersonic speeds, faster than 343 metres per second, and therefore must be able to withstand the hot temperatures and erosion caused by the gases.

The amount or surface area of propellant around the cavity that is exposed to the high temperatures directly affects the thrust produced. A higher surface area will increase thrust but reduce the burn time, as the fuel will be used up more quickly. A constant thrust requires the burning surface area to be constant. This can be achieved by different cavity core designs as can be seen in Figure 5.3. End burning has no cavity core and the propellant burns constantly like a cigarette, therefore producing a constant thrust. However, as the surface area is relatively small, only a small thrust is achieved. In star core burning the surface area of the grain remains relatively constant over time, and so it too produces a relatively constant thrust, but as the area being burnt is greater, it produces a larger thrust than the end burning. In a cylinder core grain the surface area and, therefore, the thrust, increases with time.

The grain and cavity core can be designed for different applications. For example, an initial high thrust for take-off may be required, followed by a lower thrust after the launch. Once the rocket is ignited it cannot be varied from the predetermined thrust or duration, and unless the rocket incorporates design features such as the ability to rotate or blow off the nozzle, the rocket will

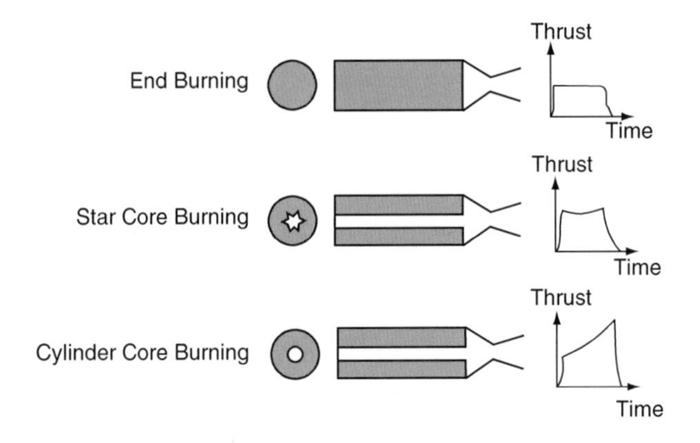

Figure 5.3 Different Core Cavities Produce Different Amounts of Thrust.

consume all of the fuel without any option for shutoff or thrust control. If the rocket is somehow shut off, it can only be lit again if a second igniter is present. Most large motors take a few seconds for the grain to ignite.

Liquid Propellants or Rocket Engines

Liquid propellant engines are more complex and less reliable than solid rocket motors. However, they have many advantages. They can be tested on the ground prior to launch and can be throttled down, stopped and restarted. They usually have the highest specific impulse or Isp rating of any form of propellant. Some liquid propellants can be stored for up to 20 years and most types produce a non-toxic exhaust. The propellants are stored in tanks and fed into a combustion chamber or rocket thrust chamber by a pressurized gas or a pump. Gas pressure-fed systems are usually used on low thrust, low total energy propulsion systems, for example, for controlling the direction the spacecraft is pointing. Pump-fed systems are used where larger amounts of propellants and higher thrusts are required, such as in launch vehicles.

The propellant storage tanks and the high-pressure gas tanks must be kept as light as possible as they also have to be lifted by the engine. They can be made from the same materials as a solid rocket motor casing, but if the tanks are made from fibre-reinforced plastics, they must contain a thin metal inner liner to stop leakage through the fibre-reinforced walls. A spherical tank contains the least shell material for internal volume. However, spherical tanks are inefficient in their use of space in the vehicle and are expensive to build and, therefore, a compromise between cost and efficiency is often required in the design. Most tanks are cylindrical with rounded ends, but they can be irregular and are often made as an integral part of the rocket's fuselage. The tanks can be placed in various locations in the rocket and are often used to move the centre of gravity, which can help to balance the spacecraft during the launch. Liquid propellant engines require controls to start and stop the engine and to restart it if required. They may also contain controls to fill and drain the propellant and various safety devices that self-check the proper functioning of the systems.

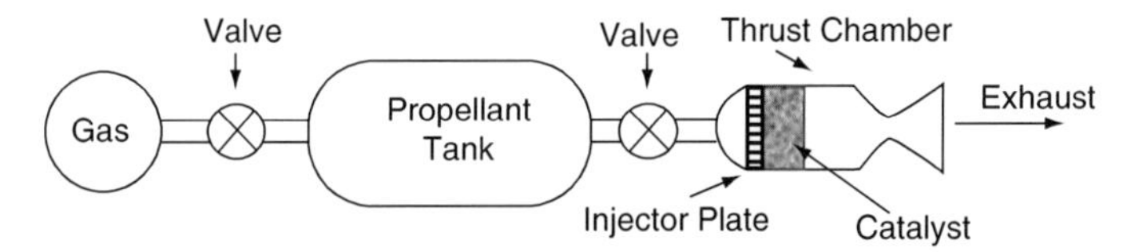

FIGURE 5.4 Pressure-Fed Monopropellant Rocket Engine.

Monopropellant

Certain liquids, such as hydrogen peroxide, also known as High Test Peroxide or HTP, can be made to form hot gases by decomposition. When HTP passes over a catalyst of a platinum or silver mesh it decomposes into steam and oxygen. The British developed sophisticated HTP motors between the 1940s and the 1970s including the Sprite, which was designed for Rocket Assisted Take-Off, and the Stentor motor, used in the *Blue Steel* air-to-ground missile. The Germans used HTP to drive the turbine for the fuel pump of the *V2* rockets. Hydrazine, produced from ammonia, is also used as a rocket fuel. A common catalyst for hydrazine is alumina granules, coated with iridium. Some monopropellants need to be ignited by an electrical heater or a flame before they decompose. Figure 5.4 shows a monopropellant rocket engine design, where the gas is pressurized and used to force the propellant into the thrust chamber.

Bi-propellant

Most liquid propellant engines are bi-propellant, that is, they use two separate liquids, a fuel and an oxidiser. The two liquids are usually stored in separate tanks and are only mixed when they are required to produce a thrust. Most fuels are hydrocarbons, such as gasoline, kerosene, diesel oil and turbojet fuel, although liquid hydrogen is also used. The storage of very low temperature propellants, called cryogenic propellants, is often not possible unless sufficient insulation is available. The most common oxidiser is liquid oxygen. As this boils at –182.9°C, it is the cold oxygen tank that is responsible for the ice that forms and falls off most large rockets during take-off. The ice is water from the atmosphere that has frozen onto the side of the tank. Other oxidisers include nitric acid. Simple pump-fed and pressure-fed liquid bi-propellant rocket engines are shown in Figure 5.5.

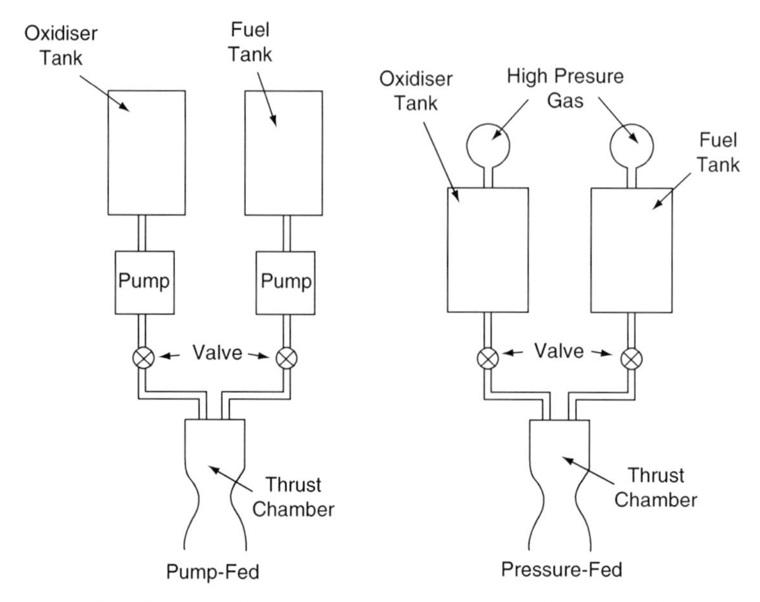

FIGURE 5.5 Simple Pump-Fed and Pressure-Fed Liquid Bi-Propellant Rocket Engines.

Liquid Propellant Engine Design

The rocket thrust chamber is the major part of the rocket engine and consists of the injector, nozzle and combustion chamber. Figure 5.6 shows a basic schematic of a thrust chamber.

The fuel and the oxidiser are forced through small holes in the injector plate into the combustion chamber. The design of the holes ensures the correct proportion of fuel to oxidiser and also breaks down the liquids into tiny droplets, a process called atomisation. Usually the holes direct the two liquids into the path of each other to enable good mixing. The gap between the injector plate and where the jets impinge insulates the plate from the heat within the chamber and helps to prevent it melting. For a monopropellant the injector plate is required to atomise the propellant before it passes over the catalyst and out through the nozzle.

The combustion chamber is where the fuel and oxidiser mix, ignite and combust. If the fuels are not hypergolic, that is they do not ignite spontaneously on contact, a source of ignition is required. The choice of material for the chamber is dependent on the fuels burnt. A copper alloy can enclose non-corrosive gas mixtures, such

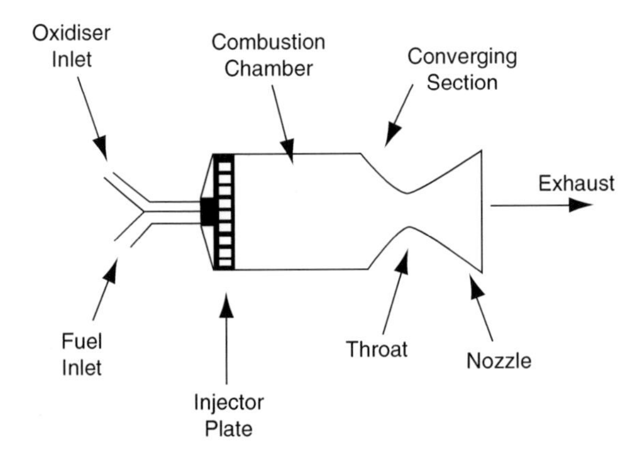

FIGURE 5.6 Schematic of a Thrust Chamber.

as hydrogen and oxygen, whereas corrosive gas mixtures, such as those that contain nitric acid, require a stainless steel chamber. As different propellants react at different rates, the chamber design is dependant on the fuel used.

The nozzle is designed to expand and accelerate the hot gases to supersonic speeds. It does this in the converging section by reducing the area the gas must flow through. Just as a constriction in a river makes the water flow faster, the converging section of the nozzle makes the gas flow faster. However, unlike the water in the river, gases can be compressed and expanded. The compressibility of a gas can easily be demonstrated with the aid of a bicycle tyre pump. If you put your thumb over the end of the pump and push the handle, the air inside the pump will squash. It is quite easy to squash it to half the initial volume, and, with a bit more effort, to less than this. The air is very springy and if you let go of the pump handle, it will expand again and go back to almost where it started. With careful nozzle design, when the gas passes through the throat it will have reached the speed of sound or become supersonic. The speed of sound is how fast the pressure disturbance, also known as sound, moves through something, such as a gas. It does this by particle interaction. As one particle moves, it exerts a force on its neighbouring particles, thus disturbing those particles, which in turn move their neighbours, like a Mexican wave. The sound energy is therefore transported through the medium. The speed of a sound wave is how fast the disturbance is passed from particle to particle. Therefore, if a gas is travelling faster than the speed of sound, pressure disturbances cannot keep up.

When the exhaust gases pass through the expansion part of the nozzle, it causes a reduction in pressure. If the fluid has not reached the speed of sound, this pressure reduction or disturbance would propagate backwards and cause the gas further back along the nozzle to accelerate. However, if the gas in the nozzle is supersonic, this pressure disturbance cannot travel backwards as the gas is moving too fast and the energy produced must be dissipated in another way. It is therefore transformed into kinetic energy, which further accelerates the gas in the nozzle. If the gas had not reached a supersonic speed by the time it reached the nozzle, its speed would decrease, just as a wider river flows more slowly. If the gas passed the sound barrier before reaching the throat, the narrowing walls would force it to slow down.

Cooling Systems

The temperature within the chamber and the nozzle can reach between 2,500°C and 4,100°C. This is much higher than the melting point of most of the materials that could be used to make the chamber and nozzle. Therefore, the walls must either be cooled or the engine switched off before they become too hot. As switching the engine off severely limits the engine or motor, two other methods of cooling are usually used. The first is to ensure the temperature of the chamber does not exceed a maximum temperature. A process that cools the thrust chamber and preheats the propellant is called regenerative cooling. One of the liquid propellants is circulated around a cooling jacket that is wrapped around the thrust chamber. The thrust chamber is therefore cooled while the propellant is warmed, which means more energy can be extracted from it. This type of cooling is usually used in bipropellant chambers of medium to large thrust engines. Figure 5.7 shows a schematic of a regenerative cooling system for a thrust chamber.

The other method of cooling is to let the chamber temperature continually increase, but reduce the wall temperature by absorbing the heat in an inner liner of ablative material. As the liner heats up it chars or melts and is finally lost by vaporization. Char has a

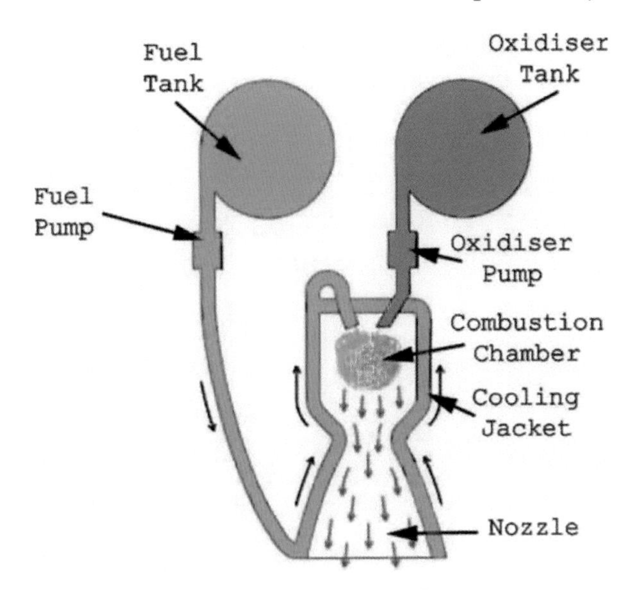

FIGURE 5.7 Regenerative Thrust Chamber Cooling.

poor thermal conductivity as the heat does not pass through it very well, so it protects the surface of the walls. The gases and vapour given off during the process also take the heat away from the surfaces. This type of cooling is used extensively on solid propellant rocket motors.

To cool monopropellant engines, radiation cooling is often used. The chamber and nozzle are made of a single wall of a material that can withstand a high temperature. When it glows red or white hot, the heat is radiated away to the surroundings or to empty space. Other methods of cooling are sometimes used including film cooling and special insulation, although these are not covered in this book.

Propellant Feed Systems

The propellants are fed into the combustion chamber either by using gas pressure or by pumps. The gas, which is often nitrogen, is stored at high pressure and, by using valves, it is allowed to enter the propellant tanks, where it forces the propellant into the combustion chamber. In large rockets where the propellant must be at a high flow rate, the gas pressure would have to be very high and the chamber containing the pressurized gas would have to be very strong and heavy. Pumps are, therefore, usually used to feed large liquid propellant rockets. The power to run the pumps is often supplied by a turbine driven by a gas generator. Hot gases exhausted from a small combustion chamber turn the blades of a turbine, which in turn rotates gears, which powers the pumps. The exhaust gases from the turbine travel relatively slowly and can either be added into the main combustion chamber or exited through a separate exhaust nozzle. Either way, they add a small amount to the thrust of the engine. Usually the main rocket propellant can be used for the gas generator, but sometimes a special propellant, such as hydrogen peroxide, is used. All feed systems also require piping, valves, a method for filling and draining and control devices to stop, start and regulate flows. This all adds to the mass of the rocket. Gas pressure-fed systems are usually used on low thrust, low total energy propulsion systems, as they are more lightweight than pump systems and they work well for repeated short duration pulses of thrust.

Gelled Propellants

Some liquid propellants have additives that turn them to the consistency of jelly when they are still, but return to a fluid when agitated. Ketchup and non-drip paint have similar properties and materials like this are called thixotropic. The gelled propellant can flow through pipes, valves and pumps when an adequate force is applied to it but return to a solidified gel when there is no force. Therefore, there is less chance of a spill or leak, which can be hazardous, corrosive or toxic. Sloshing in the fuel tanks, which can cause flight instability, is also eliminated. However, the loading or unloading of gelled propellants is more complex than for normal liquid propellants, and gelled propellants also have a slightly lower Isp, which makes the fuel less efficient than liquid propellants.

Hybrid Propellants

A hybrid propulsion system is where one propellant is stored in a liquid phase and the other as a solid. Most often the oxidiser is a liquid and the fuel a solid. Like a solid rocket motor, the fuel is packed into a grain. The pressurized oxidiser is forced through the grain causing it to vaporize and the resulting exhaust enters a mixing chamber to ensure complete combustion before exiting through a nozzle. This type of system has the advantages of safety during fabrication, storage and operation, stop, start and restart capabilities and a higher Isp than solid rocket motors. By reducing the flow of the liquid oxidiser, the mixture ratio and, therefore, the thrust can be varied. The hybrid-engine powered *SpaceShipOne*, which won the Ansari X-Prize, burnt nitrous oxide, more commonly known as laughing gas, as the liquid oxidiser and hydroxyl-terminated polybutadiene (HTPB) rubber, which is also used in car tyres, as the solid fuel.

Cold Gas Propellants

Cold gas propellants are very simple, usually consisting of a high-pressure gas tank, valves and simple nozzles. This system is similar to a deflating toy balloon flying around the room, although the escaping gas is controlled and directed through the appropriate nozzle to produce the required action and reaction. The gases that are often used are often high density gases and include nitrogen, argon and krypton.

Electric Propulsion

There are three main types of electric propulsion, classified according to the method of accelerating the propellant. These are electrothermal, electrostatic and electromagnetic. In all of the electric propulsion systems an electric power source is required, but this is usually separate to the propulsion system. Electric power sources are described in Chapter 2 – "Rockets and Spacecraft". The thrust from an electric propulsion system is usually low, about 0.005–1 newton, and therefore only a small acceleration is attained, although this acceleration can be sustained for a period of weeks or months and therefore has a very high Isp rating.

Electrothermal Systems

In an electrothermal system, the propellant is heated electrically and then expanded through a nozzle, in the same way as a chemical rocket. The propellant is usually a gas, such as hydrogen, ammonium or nitrogen. There are two basic methods of heating the propellant, either resistojets or arcjets.

Electric Resistojets

These are the simplest form of electric propulsion. The propellant flows over heated elements, in a similar way to an electric hot water heater or shower. The propellant is then accelerated through a nozzle. The melting point of the wall materials limits the temperature this design can be used for. Resistojets are used on the *Iridium* communication satellites, with hydrazine as the propellant.

Arcjets

Arcjets heat the propellant by passing an electric arc through it, in the same way that a car uses a spark plug to heat and ignite the petrol and air mixture. The amateur radio satellite *AO 40*, launched in November 2000, actually used a car's spark plug in its arcjet thrusters. This method overcomes the resistojet's temperature limitations, as the temperatures in the propellant stream can get higher than the wall's melting temperature. The arc does not usually heat all of the propellant, and therefore mixing is required to distribute the heat.

Electrostatic Engines or Ion Drives

To understand this type of engine, a basic knowledge of nuclear physics is required, although this is not as complicated as it may sound. Everything is made up of atoms. Atoms contain particles that have no charge, called neutrons. They also contain electrically charged particles, which are either positive or negative. The negatively charged particles are called electrons and the positively charged particles are called protons. When there are the same amount of electrons and protons, the item is electrically neutral. Movement of electrons is called electricity. If the electrons do not flow but gather in one place, it is called static electricity. When an electrically charged particle comes near another particle with the same charge, they repel each other. However, when a positively charged particle comes near a negatively charged particle, they are attracted to each other. This can be seen when a balloon is rubbed on a jumper. Electrons move between the balloon and the jumper, giving each a slight charge. If the balloon is then held against a wall it is attracted to it, and it stays stuck to the wall until the charge dissipates. When electrons have been removed from an atom, the atom becomes a positively charged ion. These positive ions are attracted not only to electrons, but also to a negatively charged electric field. The higher the potential or strength of the field, the faster the positive ions will accelerate towards it.

The electrostatic, or ion propulsion engine, makes use of the attraction between positive ions and a negative electric field. The propellant, usually a gas such as xenon, is ionized to become mainly positively charged ions of gas. The ions are attracted towards a static electric field that is generated by a grid with a negative electrical charge. As the ionized gas particles are a lot smaller than the holes in the grid most of them pass straight through it and are released into space, thus accelerating the spacecraft. A jet of electrons is shot into the exhaust to make sure that the spacecraft and the exhaust gas remain electrically neutral. The amount of gas flowing through an electrostatic engine is small compared to the amount flowing through a chemical engine, however, as it is accelerated to a very high speed, it has a high Isp rating. The thrust is smaller for ion propulsion engines than for chemical engines, so it is more suitable for stabilizing satellites and for providing a very slow means of acceleration, which can be used to eventually attain a high velocity or a different orbit. Chemical engines are limited by how much propellant they can carry and after a time will run out of fuel and not be able to accelerate further. Ion propulsion engines are usually only limited by their energy supply. On long journeys, such as interplanetary missions, ion drives can accelerate for much longer than chemical engines and therefore the spacecraft can travel faster and reach its destination quicker.

Electromagnetic or Magnetoplasma Engines

Electromagnetic engines or magnetoplasma engines use the properties of plasma, electricity and magnetic fields to accelerate the propellant. A plasma is a gas that has had most of its molecules split into ions, electrons and neutral particles. This type of gas is used in plasma televisions, fluorescent tubes and also enables lightning to strike.

Lightning strikes happen when the air has been changed from a gas to a plasma. The bottom part of a thundercloud has a negative electric charge. This repels the electrons in the Earth deeper underground, which leaves the surface with a positive charge. The difference in electric charge between the cloud and the Earth produces an electric field, which ionizes the air and turns it into a plasma. The charged particles in the plasma are free to move and so they can conduct electricity. Therefore an arc of electricity, more commonly called a bolt of lightning, can travel from the cloud to the Earth, just as if a metal rod connected the two. Even with the wind moving the air, the bolt will be almost vertical.

A magnetoplasma engine uses a plasma so that a current can be conducted across it. This produces a magnetic field perpendicular to the current. The magnetic field is sometimes enhanced with magnets. When an electric current and a magnetic field are at right angles to each other, a force perpendicular to both of these is produced. This force can move the medium conducting the current. To imagine these three directions, look into the corner of a room. If the current goes from the floor to the ceiling, and the magnetic field goes along the floor to the right, then the force will be directed along the floor to the left. This principle is used in loudspeakers and electric bells and the force is called the Lorentz force. Therefore, in a magnetoplasma engine, when a current is passed through the plasma and the magnetic field is produced, it also produces a Lorentz force. This force pushes the plasma out into space and causes the spacecraft to be propelled.

The main types of electromagnetic thrusters are called Magneto-Plasma-Dynamic (MPD), and Pulsed-Plasma Thruster (PPT). The MPD is a combination of an electrothermal arcjet and the magnetoplasma engine, where both heating and the Lorentz force accelerate the plasma. As its name suggests, the PPT does not run continuously, but pulses for the duration of the charge on a capacitor, and then must wait until the capacitor is recharged.

Nuclear Propulsion

The only use of nuclear energy in space so far has been to provide electricity. The theory of the use of nuclear power for other propulsion system has been developed and experimental tests have been conducted on the Earth. Further information on these systems is included in Chapter 11 – "The Future".

Nuclear Electric Rocket

Electrical power from either radioisotope thermoelectric generators or nuclear fission reactors, as described in Chapter 2 – "Rockets and Spacecraft", can be used to power ion drives. Only one American spacecraft has been powered using nuclear fission. As mentioned previously, it was the *Space Nuclear Auxiliary Power* or *SNAP-10A* reactor, launched in April 1965. This used uranium 235 as a fuel. The reactor was used to charge the batteries for an ion engine. However, the ion engine operated for less than an hour before being shutdown. This was due to a fault in the spacecraft and the reactor successfully worked for 43 days. The spacecraft is still in orbit around the Earth.

Solar Thermal Propulsion (STP)

Propulsion systems that heat and expand propellants usually require combustion or electric power. However, the heat from the Sun could be used directly. By focusing the sunlight onto a chamber containing the propellant, the propellant would be heated naturally, in the same way as the air inside a greenhouse is warmed on a hot summer's day. Research has shown that it is feasible to transfer a satellite from LEO to GEO using this method and it could also be used for exploration of the inner solar system. However, no STP system has yet been used in space as the main source of propulsion, mainly because the current STP systems, with their propellant, are larger in volume than the technologies already in use. They therefore do not fit into the space on the launch rockets where the chemical or electrical propulsion systems are currently located. The benefit of the STP being a lighter system does not yet overcome this disadvantage.

Other Forms of Propulsion

Propulsion systems based on the reaction motor have been used almost exclusively for space flight. This has limited the speed and distances we have been able to travel, as the fuel or propellant must always be carried with the spacecraft. Other forms of propulsion have been suggested and some have become technically and financially viable, such as solar sails, which use sunlight to propel the craft along, but others, such as space elevators, that use a solid link between the Earth and a geostationary orbit, remain, for now at least, in the realms of science fiction due to technical constraints. Possible non-reaction motor propulsion systems of the future, including solar sails and space elevators, are described in Chapter 11 – "The Future".

6. Navigation in Three Dimensions

On the Earth we usually travel on the surface of the land. Even if we go up and down hills we perceive our journey as being relatively flat and so we can draw our route on a map in just two dimensions. The compass bearings of north, south, east and west are usually used to show the directions. When we describe the location of somewhere on the Earth, we often only have to give a distance and direction from some other known place. For example, Manchester in the UK is about 300 kilometres north of London. We ignore that they may be at different heights above sea level and that the surface of the Earth is curved. This type of description is known as a local coordinate system and as long as the starting place, distances and directions are well defined, anyone can use it.

Navigating in space is more complicated, as the third dimension of up and down, which we can usually ignore on the Earth, becomes more important. Also, we must choose a reference system that is relevant to the scale of the mission. On Earth, all locations are stationary relative to one another. Manchester was about 300 kilometres north of London 1,000 years ago and will be the same distance away in 1,000 years time. In space however, all things move. Spacecraft and the Moon move relative to the Earth, the Earth and the planets move relative to the Sun. Even the Sun moves relative to our galaxy the Milky Way, and the Milky Way moves relative to other galaxies. However, the movements of the Sun and Milky Way do not need to be considered unless we are locating spacecraft outside of the solar system or we are considering timescales of hundreds or thousands of years. For spacecraft, the reference system could be the Earth, the Sun, the spacecraft itself or even another planet or object that the spacecraft is approaching.

Vehicles on the ground on the Earth generally have a front and back at opposite ends of the length of the vehicle. These vehicles then usually move in the direction they are pointing. However, at sea or in the air, currents can move the vehicle sideways, while it is travelling forwards. The way in which a vehicle is pointing with reference to its direction of travel is called its attitude. A spacecraft may travel in the direction it is pointing when the engines are delivering thrust, but after the engines have cut-off and the vehicle is in free fall, the spacecraft is unlikely to be pointing in the direction of travel. Navigation in space is concerned with the vehicle's location in space, not its attitude.

Coordinate Systems

In 1637 the French philosopher and mathematician Rene Descartes, famous for his statement "cogito ergo sum" or "I think, therefore I am", invented the Cartesian coordinate system. This system can describe any position in space. The Cartesian coordinates divide directions into the *x*-, *y*- and *z*-axes, which are all perpendicular to each other and linear, as can be seen in Figure 6.1. Most Cartesian systems use a right-handed set. If you use your right hand, and point your thumb upwards, your index finger straight ahead and your middle finger to the left, they form a right-handed set. Most Cartesian systems use these directions for the *x*-, *y*- and *z*-axes.

The place where the zeros of the *x*-, *y*- and *z*-axes of a Cartesian system meet is called the origin. As long as the origin is known, any position can be determined. Generally, movement left or right is shown in the *x* direction, up and down is shown in the *y* direction and forwards and backwards in the *z* direction. Moving one unit anywhere on the *x*-axis, would cover the same distance as moving one unit anywhere on the *y*- or *z*-axes.

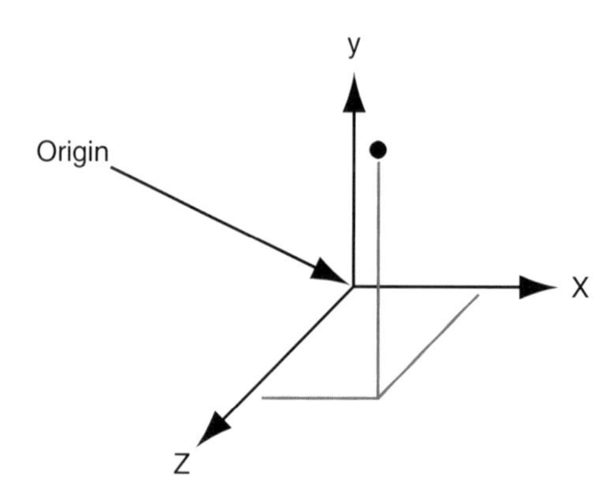

FIGURE 6.1 Cartesian Coordinate System.

A polar coordinate system is one where the relationship between two points is expressed in terms of angles and distance. If this information were required from a Cartesian system, trigonometric calculations would be needed. Each point on a two-dimensional polar coordinate system is represented by the radial coordinate, designated by the letter r, and the azimuth or polar angle, also known as the angular coordinate, usually designated by the Greek letter θ, as can be seen in Figure 6.2. The radial coordinate is the distance from the central point, known as the pole. The pole is the equivalent to the origin in the Cartesian system. The azimuth angle is the angle from the 0° ray or the polar axis, which is equivalent to the *x*-axis on the Cartesian coordinate system. Sailors and hikers often use this type of coordinate system for navigation and route planning, but they call the azimuth angle the bearing.

By adding a second distance dimension, usually labelled the *z*-axis, perpendicular to the two-dimensional polar coordinate system, a cylindrical coordinate system can be used for items in three-dimensional space, as can be seen in Figure 6.3. Air traffic controllers use this form of system, with altitude on the *z*-axis.

Alternatively, the addition of another angular dimension produces a spherical coordinate system. An imaginary flat sheet that

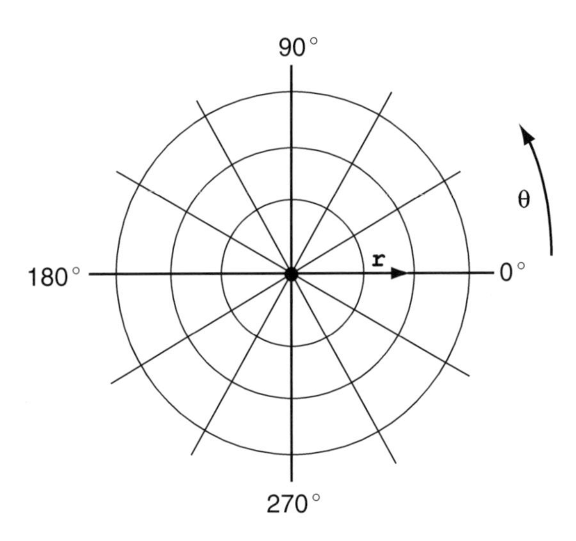

FIGURE 6.2 Two Dimensional Polar Coordinate System.

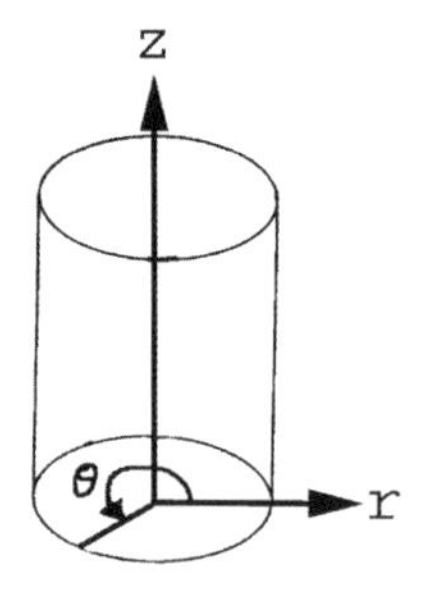

FIGURE 6.3 Cylindrical Polar Coordinate System.

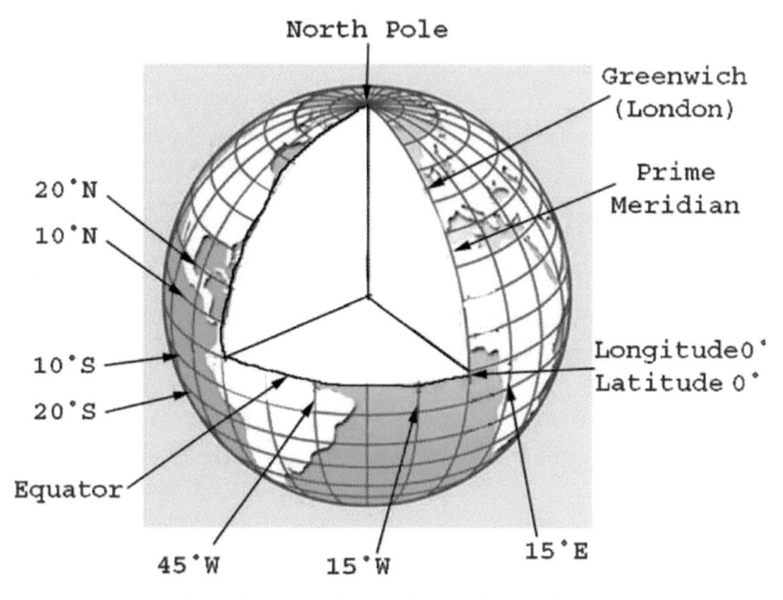

FIGURE 6.4 Longitude and Latitude – Spherical Coordinate System.

divides the sphere into two hemispheres is called the fundamental plane. The latitude and longitude system is a spherical coordinate system, with the pole or centre of the system being at the centre of the Earth as can be seen in Figure 6.4. The lines of latitude are angles above and below the equator or fundamental plane, with the lines of longitude measuring the angle from the line of the Prime Meridian, which is also known as the Greenwich Meridian or the International Meridian.

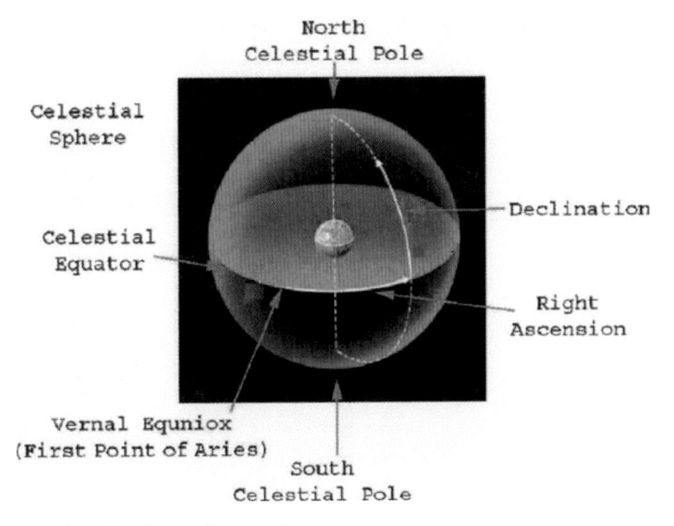

FIGURE 6.5 Celestial Coordinate System.

The celestial coordinate system is a spherical coordinate system that is used to catalogue the stars and is based on the celestial sphere, as can be seen in Figure 6.5. The celestial sphere is an imaginary sphere encompassing the Earth. All objects in the sky can be thought of as being on the sphere. The celestial equator and the celestial poles are projected from their corresponding Earthly equivalents. Angles measured north or south from the line of the celestial equator are called the declination and are measured in degrees. This is similar to lines of latitude. The longitude equivalent is called the Right Ascension or RA. This is measured in hours eastwards from the First point of Aries, which is given the symbol ♈ and is also known as the Vernal or Spring Equinox.

The Sun does not always appear directly above the equator, but follows a path called the ecliptic, which takes it above and below it. When its path crosses the equator, it is called an equinox. For the northern hemisphere, the Vernal Equinox is usually about March 20 and is when the length of daylight and darkness is exactly equal, and the Sun is directly above the equator. This marks the beginning of Spring. The word equinox comes from Latin aequi or equi, meaning equal and nox or noct meaning night. When the Sun reaches its limit north or south of the equator, it is called a solstice.

The Vernal or Spring Equinox for the northern hemisphere is when the Sun crosses the equator from south to north. The Autumnal Equinox is the beginning of autumn in the northern hemisphere and is usually about September 23. This is when the Sun crosses the celestial equator from the north to the south.

When it was first named the First Point of Aries, the Sun was in the constellation of Aries at the Vernal Equinox. Since about AD 500, however, the Sun has been in the constellation of Pisces at the Vernal Equinox. This is due to precession, caused because the Earth wobbles. The Earth's axis, around which it spins, does not point straight upwards through its orbital plane around the Sun. This orbital plane is an imaginary sheet on which the Earth's path around the Sun would lay. It is because the axis is slightly skew to the orbital plane we get the four seasons of spring, summer, autumn and winter. Over about 25,800 years, the Earth's axis traces a circle in space, just like a spinning top wobbling about. This is called precession and means that over time, the stars appear in different places. Due to this movement, it is necessary to define the time when any coordinate system refers to. By internationally accepted convention, 12:00 on January 1, 2000, is normally used nowadays, referred to as the epoch J2000.0. Before this epoch was used, for astronomy and space travel, the epoch B1950.0 was used.

Star positions are known very accurately using the celestial coordinate system, and so a photograph of an Earth orbiting satellite against a star background can be used to calculate the satellite's Right Ascension and declination at the time of the photograph. As most stars are so far away, they have almost the same coordinates from everywhere on the Earth's orbit around the Sun and this system can be used at any time of the year.

The type of coordinate system used when defining the location of a spacecraft depends on where the spacecraft is in relation to the Earth and why the location is required. Usually, a Cartesian system is used, with the origin located either at the centre of the Earth, or the centre of the Sun. However, other origins, such as the centre of gravity of the spacecraft, can also be used. As long as the system is well defined, any reference given in a particular coordinate system can be transferred to any other coordinate system. Some of the different types of coordinate systems used for locating spacecraft are included in Appendix B – "Coordinate Systems".

Locating Spacecraft

To find a particular spacecraft at a particular time, three measurements are required to determine its current position in whichever coordinate system is used. To determine the path of the satellite the speed of the spacecraft in each of the three directions is also required.

During free fall flight, a spacecraft is not guided and may tumble randomly about its centre of mass. If the spacecraft were on the surface of the Earth, this would be the same point as its centre of gravity. If the path is not perturbed by the influence of anything else, such as the gravity of the Sun or the Moon, the path of the centre of mass would be a smooth track in space, but any point on the craft, such as instruments or solar panels, may spin. If they are required to point in a particular direction, the satellite will need to be stabilized, as described in Chapter 3 – "Rockets and Spacecraft".

During the free fall part of a spacecraft's flight, the position, direction and speed of the spacecraft are relatively predictable. The speed and direction are usually combined into one measurement called the velocity.

If the gravitational field was precisely known, then starting with the vehicle's velocity and its position, the entire free flight trajectory could be calculated. These are intricate calculations and are usually solved with the aid of a computer. However, as even the Earth's gravitational field is not precisely known everywhere, the calculated trajectory will not be exactly the same as the actual trajectory. These gravity differences are caused by irregularities in the Earth, such as it not being perfectly spherical or of a uniform density. As the Earth bulges at the equator, at the surface the gravity is slightly weaker in the equatorial regions than it is at the poles. Also, the local topography, including mountains, and geological conditions, such as the granite found in Edinburgh, Scotland or the chalk in Dover, England, affect the gravitational field by a small amount. Other small forces may change the path of a satellite slightly, including meteoroid impact, magnetic fields, solar wind and electromagnetic radiation. If more than two bodies are involved, for example, the Sun, the Moon and the Earth, the path cannot be accurately predicted in the long term, but for practical purposes and for relatively short-term space missions, this does not matter too much. The accurate location of spacecraft is determined by calculation and other positional information.

Inertial Navigation

Inertial navigation is when the course of a vehicle is calculated without the use of external observations or equipment. It is a form of dead reckoning that has been used in one form or another by travellers throughout time. For example, onboard a ship, the speed, direction and time travelled can be used to calculate the distance travelled and so the estimated current position can be plotted.

The basic form of inertial navigation used in space consists of accelerometers and gyroscopes. Accelerometers are instruments that measure acceleration and so allow the speed to be calculated. They are used to measure the acceleration in the three straight line degrees of freedom, up and down, left and right and forward and backwards. The simplest accelerometers consist of a mass suspended between two springs. When the device is at rest or is in motion at a constant speed, the mass will stay at a neutral position. However, as the device is accelerated or decelerated, the mass will appear to move, due to inertia. By connecting the mass to a potentiometer, a device that will give a discrete voltage reading for each position of the mass, the acceleration can be determined.

Other accelerometers use piezoelectric crystals instead of a mass. These crystals acquire an electric charge when compressed and the voltage generated can be measured and related to the acceleration. There are many other ways of measuring the acceleration, including using capacitance, light and even hot air bubbles.

Three linear accelerometers positioned at right angles to each other are used to measure acceleration in each direction. To ensure that the accelerometers are always pointing in the same direction, they are suspended in a set of three gimbals that are stabilized by gyroscopes. This is known as a gimballed platform. A spinning gyroscope resists changes in the direction of its axis. By using sensors on the gyroscopes to detect any rotational movement and motors on the gimbals, the stable platform remains at a constant orientation to inertial space. By measuring the amount of movement in the gimbals, the roll, pitch and yaw of the spacecraft can be determined. This type of inertial navigation system can be extremely reliable and accurate. However, they are mechanically very complex and can be expensive and time-consuming to maintain.

In the 1970s, the inertial navigation industry started to investigate a simpler system, one without gimbals. This was called a strapdown system. To do this with conventional spinning wheel gyroscopes was very difficult. The main breakthrough came in the form of a Ring Laser Gyroscope (RLG). This is a solid glass block, with three narrow tubes drilled in it, forming a triangle. A mirror is placed at each corner of the triangle. The tubes are filled with a mixture of helium and neon. A beam of laser light is generated and split into two beams of light, one going clockwise around the triangle, bouncing off the mirrors, the other anticlockwise, as can be seen in Figure 6.6.

When the gyroscope is at rest, the two beams will travel the same distance around the gyroscope in the same time. If one of the mirrors is partially transparent, samples of both beams can be taken at the readout sensor. If the block rotates in a clockwise direction, a photon in the beam going clockwise will have

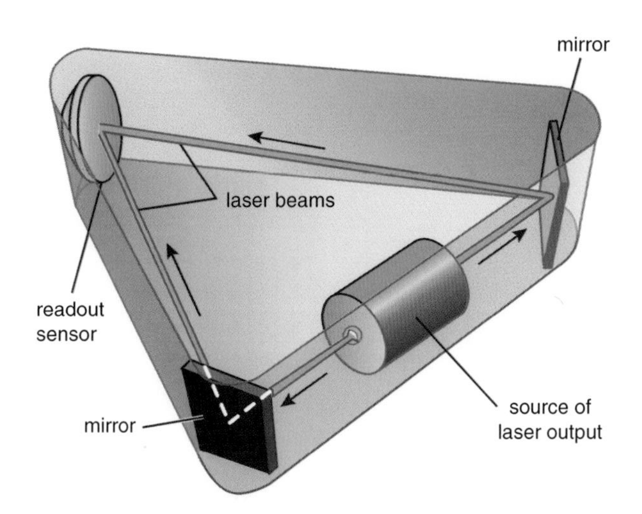

FIGURE 6.6 Ring Laser Gyroscope.

to travel a little further as the readout sensor has moved away. Similarly, a photon in the anticlockwise beam will travel less as the readout sensor has moved closer. At the readout sensor, there will then by a small difference in the two beams. There will be a change in frequency and also a phase shift between them. Where the beams cross they will interfere with each other like ripples from two stones dropped into a pond. By observing this interference pattern of the beams, the difference in path lengths can be measured and the rotation rate determined. This type of gyroscope is smaller, weighs less and is also cheaper than conventional gyroscopes and they were used in the *Ariane 4* launch rockets. In theory, RLGs should have no moving parts. However, small dither motors are incorporated to overcome an effect known as lock-in. This is when the frequencies of the two laser beams lock together and the interference pattern will not move relative to the movement of the RLG and so, over time, it will not track correctly. The dither motors make the RLG oscillate or shake quickly, which stops the laser beams locking together.

Another type of gyroscope, called a Hemispherical Resonator Gyro (HRG) is also used. This is a very small sensor that does not have any moving parts. It is shaped into a hemispherical shell and made of quartz, and looks somewhat similar to a brandy snifter glass. A standing wave is generated within the hemisphere, similar to a wineglass "singing" as you slide your finger around the rim. Electrodes within the shell can determine how much the standing waves move when the unit is rotated. These have been used on the *Near Earth Asteroid Rendezvous (NEAR/Shoemaker)* spacecraft and the *Cassini* spacecraft.

Magnetometers

A magnetometer measures the intensity and direction of a magnetic field. This type of sensor is lightweight, simple and reliable and can be used for attitude determination. However, as the Earth's magnetic field is not precisely known as it has anomalies and shifts with time, this method is not very precise. Also, the Earth's magnetic field decreases with altitude, and so a magnetometer can only be used reliably in a low Earth orbit.

External References

Even with the best instruments and highest calibration, cumulative errors add up with inertial navigation systems. These can be caused by many sources, including drift error and instrument misalignment. Therefore, they must occasionally be realigned with external data. The external reference can be obtained onboard, such as using a celestial or star tracking system, which uses the Sun or a particular star to determine the position, or an Earth-based system such as radar or optical tracking or even a global positioning system (GPS).

Sun Sensors

The Sun has been used as an aid to navigation for many thousands of years. Its brightness dominates the Earth's sky and, as it is such an unambiguous reference point, it continues to be used for navigation in space. Sensors that detect the Sun are not just used for navigation. They also perform functions and are used to determine the orientation of scientific instruments, detect the spin rate and control switching and timing operations. Different types of sensors are used for different purposes. Presence detectors only monitor whether the Sun is or is not in the field of view. These are used to position the spacecraft or instruments either to protect them from the Sun's radiation, or to enable the solar energy to be harnessed. Analogue Sun sensors provide a continuous output. This is usually a function of the angle of incidence, or angle that the light from the Sun hits a photocell in the instrument. Combined with the results from other photocells, the direction of the Sun in relation to the spacecraft can be determined. Digital sensors consist of two parts, a command unit that is basically a Sun presence detector and a measurement unit that provides a digital output that represents the angle of incidence of the Sun relative to the sensor face. From these, the angle of the Sun from the spacecraft can be determined.

Star Sensors and Astronavigation

On the Earth, navigation by the stars is relatively simple. In the northern hemisphere, to find your latitude, all you need to do is measure the height of the Pole star, Polaris, above the horizon. If Polaris is 90° above the horizon, you are 90° North of the equator or at the North Pole, whereas if Polaris is exactly on the horizon you are on the equator. Longitude is a little more complicated, but with an accurate clock and an up to date almanac of the positions of the stars for every day of the year, it can be found. However, in space there is no definite horizon and therefore no quick and easy method of determining the location.

Star sensors measure the direction of a star from a spacecraft. By comparing this to the direction tabulated in a star catalogue, the orientation or attitude of the spacecraft can be calculated. In the past, unexpected events, such as sunlight reflecting off a fleck of paint, giving the impression of a bright star, would sometimes make a sensor provide spurious results, however, with improvements in computer imaging and processing power, star sensors are becoming quicker and more able to cope with such events.

For space vehicles travelling to the Moon or planets in our solar system, the star Canopus is almost always used. NASA's *Lunar Orbiter* missions used astronavigation based on Canopus, until the spacecraft approached the Moon and the star became concealed. Inertial navigation was then used. Canopus is the second brightest star in the sky, excluding the Sun. This makes it relatively easy to detect and distinguish from other stars. Sirius, the brightest star, is about twice as bright as Canopus, but it has other relatively bright stars around it, which makes it more difficult to discriminate. When only one star is used, only the attitude of the vehicle can be calculated. However, combined with two other reference points, such as the Sun and the Earth, the location of the spacecraft can be determined. Ideally the angle between the Sun and the star should be near to 90°. Polaris is nearest to this angle, but it is only about a tenth as bright as Canopus and therefore not as easy to detect. As Sirius is quite close to the Sun, Canopus is in a better position for use as a reference coordinate with the Sun.

Canopus is located in the constellation Carina, and can be seen at some time of year from all locations on the Earth south of latitude 38° North. It is a supergiant and is magnitude –0.72. Carina is one of the constellations that formed part of a huge group of stars called Argo Navis, named for the ship that carried Jason and his 49 Argonauts on the quest for the Golden Fleece. Argo Navis is no longer recognised as a constellation as it has been broken into four smaller constellations. Canopus was said to be the rudder of the ship and thus guided it across the sky, just as spacecraft are currently guided by it. Canopus was also the name of Menelaus's pilot, who led the King's troops home to Greece from Troy.

Earth or Horizon Sensors

An Earth sensor is usually a thermal device that works in the infrared. They are often called horizon sensors, as they detect the difference between the Earth and space, not just whether the Earth is in the field of view. From a low Earth orbit, the Earth takes up such a large portion of the sky that an Earth presence detector would not be much use. The visible light spectrum cannot be used, as the reflected sunlight from the Earth, known as the albedo, is very bright during the day and almost non-existent during the night, with a fuzzy blurred region where the two meet. This would make sensing visible light impractical. However, the Earth emits an almost uniform glow from its upper atmosphere in the infrared, both day and night. Compared with the cold of space, the boundary between Earth and space that occurs at the horizon is easy to identify. From the detection of the horizon by different sensors, or by timing measurements of the horizon passing the field of view in a rotating spacecraft, the direction to the centre of the Earth can be determined, along with the altitude of the spacecraft. Earth sensors are generally less accurate than Sun or star trackers, but they are more tolerant to space radiation and also much cheaper.

Navigation by Asteroids

NASA's *Deep Space 1* mission, launched in 1998, experimented with a system called Autonav. This takes images of known bright asteroids. The inner solar system asteroids move in relation to stars and other bodies at a noticeable and predictable speed. By tracking them, the relative location of the spacecraft can be determined. By imaging two or more asteroids, a more exact position can be determined and by acquiring two or more positions over time, the trajectory can be calculated. The software for this analysis was carried onboard *Deep Space 1* and so the spacecraft was able determine its position without any Earth-based guidance. Combined with thrusters for steering, the Autonav system enables autonomous navigation of spacecraft and reduces cost and bandwidth requirements for certain missions. It can also be used to precisely locate asteroids, comets or other objects in space in relation to itself. The Autonav system was used for this purpose by NASA's *Deep Impact* probe in 2005. It was used to align the spacecraft with the path of comet Tempel 1, so that they collided and information could be gathered from the inside of the comet.

Global Navigation Satellite Systems or GNSS

There are currently only two working global navigation satellite systems, the USA's Global Positioning System or GPS and the Russian Global Navigation Satellite System, GLONASS. Europe's Galileo system will also provide another GNSS when it becomes fully operational, hopefully in 2012, and will provide a highly accurate, guaranteed global positioning service under civilian control. It will also be able to operate with the GPS and GLONASS systems.

The GPS satellites can provide the latitude, longitude and altitude day or night to users on the Earth and to satellites in lower orbits. They can also provide the direction and speed of travel and therefore the attitude of the vehicle can also be determined. By having three antennas attached to the GPS receiver, and measuring the difference in phase of the signals received, the angle between the vehicle and the GPS satellite can also be determined.

The GPS constellation of satellites is about 20,300 kilometres above the Earth. Other satellites, in a lower orbit than the GPS constellation, can take advantage of the GPS signals to determine their position in a similar way to Earth-based users. The GPS receiver on a satellite uses very little onboard resources, such as power and space, yet it is capable of locating the satellite to within 200 metres. The use of onboard processing can improve the accuracy significantly. Portugal's first satellite, *PoSAT-1*, launched in 1993, uses a GPS receiver in orbit. It can determine its orbit autonomously, rather than relying on ground-based post-processing which earlier satellites had to use. This saves a lot of time, money and waveband usage. NASA currently uses its ground-based Global Differential

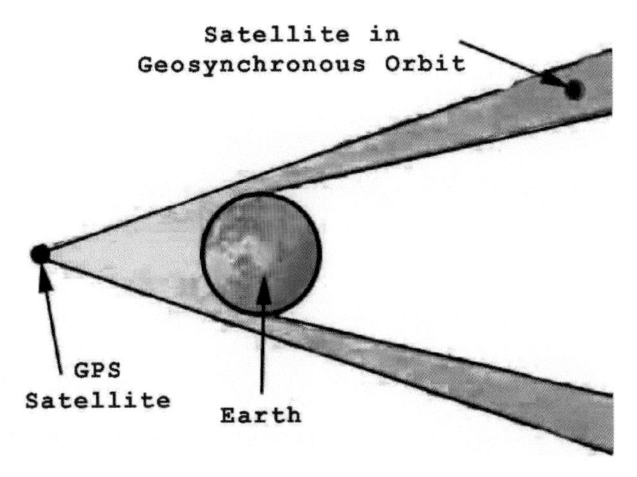

FIGURE 6.7 A Satellite in Geosynchronous Orbit can use the Signal From a Global Positioning System to Determine its Position.

GPS System for rapid and precise orbit determination for missions that require very precise real time positioning.

As each GPS satellite broadcasts a beam slightly wider than the Earth, a geosynchronous satellite situated on the other side of the Earth is also able to receive the signals, as can be seen in Figure 6.7. By interpreting the data, the satellite can then determine its position. Traditionally, operators of geosynchronous satellites have used a tracking station on the Earth to provide the range measurements to determine the orbit. As a geostationary satellite, by definition, has little relative motion with respect to the Earth's surface, orbit determination by land-based systems can be problematic. More stations spread across the Earth would provide a better system, but various constraints, including cost and available space on wavebands, make this difficult to achieve. The GPS is therefore an improvement, although as the geostationary satellite will not often be able to view four GPS satellites at once, it will not be able to discern its position in the same manner as GPS receivers on the Earth. However, as traditionally only one ground station is used, an absence of more GPS signals is not a problem.

Navigation in Manned Spacecraft

Crew operated sighting devices have been used by US astronauts since the *Apollo* space flights. The Crew Operated Alignment Sight

(COAS) is a collimator, similar to an aircraft gunsight. A reticle or cross-hair is projected onto a clear glass optical element, called a combiner or combining glass. Head-up displays on modern motor car windscreens or in aircraft use a similar system. The display is focussed at infinity, so the astronaut's eye does not have to refocus between the reticle and the object being sighted. On the Space Shuttle, a COAS is mounted at the commander's station, and is used to check the correct attitude orientation during the ascent and de-orbit. Once in orbit, the COAS is moved next to the aft flight deck and is used to check the alignment of the payload bay doors and for measuring range and rotational rates.

If the inertial measurements are out by more than 1.4°, the star trackers are unable to acquire and track stars. The COAS must then be used to realign the measurements to within this limit, after which the star trackers can be used to realign the inertial measurements more precisely. To realign the inertial measurement units, the Space Shuttle is moved so the selected star is in the field of view. When the star crosses the reticle, a button is pressed and the gimbal angles of the inertial measurement units are taken. This is repeated for another star. As the location of the stars are known, and the location of the COAS within the Shuttle is known, the inertial measurement units can be reset.

Earth-based Navigation and Tracking Systems

Almost everything in space that is larger than a grapefruit is tracked by at least one of the many Earth-based systems. Tracking objects from the ground, including, optical, radar, infrared and laser tracking are covered in more detail in Chapter 9 – "Observing Satellites".

Lunar and Interplanetary Systems

Further visits to the Moon and to Mars are currently being planned. A positioning and navigation system covering each of these bodies would be of great benefit to orbiting spacecraft, descent and ascent vehicles and also during surface activities. An integrated

Interplanetary Navigation and Communication System is being investigated, which would be similar to the Earth's GPS. The interplanetary system would provide better navigation capabilities and more precise location and target information. It would also provide an improved communication capability, enabling easier and quicker transfer of both scientific and operational data.

Time Dilation and Relativistic Effects

Most, if not all, satellites carry an accurate clock that is synchronised with an Earth-based clock. This is used for many things, including digital processing and determination of the location of the spacecraft. However, no matter how accurate the clocks are, after a while there will be a discrepancy between those on the Earth and those in space. This is because time passes more slowly if you are travelling very fast. Unfortunately, this is not just a case of "if you're having a really nice time, ten minutes seems to go by in an instant, whereas if you are waiting for something special, ten minutes can drag forever." It is a bit more complicated. Sufficient to say that in 1905, Albert Einstein showed in his Special Theory of Relativity that time runs slower during very fast movements. This is known as time dilation and was proved empirically in October 1971, when four atomic clocks were flown around the world. Compared to identical clocks that stayed at the US Naval Observatory, the clocks on the aeroplane showed a different time when they got back. A clock on a satellite travelling around the Earth at about 3.8 kilometres per second, which is the speed of the GPS constellation of satellites, runs slower when viewed from the Earth. This leads to an inaccuracy of time of approximately 7.2 microseconds or 0.0000072 seconds a day, or 0.0026 seconds a year. This does not sound much, but the GPS timing signal is typically accurate to ten nanoseconds or 0.01 microseconds, which is 0.00000001 seconds.

Eleven years after his Special Theory of Relativity, Albert Einstein showed in his General Theory of Relativity that time moves slower in a stronger gravitational field. Therefore, as a satellite circling the Earth is in a much weaker gravitational field than it would be at the Earth's surface, an onboard clock would run faster than one on the Earth. NASA has proved this with experiments in space. There is a third relativistic effect, called the Sagnac effect. This is caused by the rotation of the Earth and is dependent on the direction and the path travelled.

Due to all these different factors, relativity needs to be taken into account when synchronising clocks on the Earth with clocks on satellites, and also between clocks on different satellites. In most orbits, all three relativistic effects will change as the satellite goes around its orbit, as the height above the Earth, speed and direction of travel in relation to a ground station, change throughout the path. However, in most cases, the clocks do not have to be constantly changed, as the time dilation will be nearly the same for one orbit as the next. By setting the frequency and, therefore, the tick rate of the space-bound clocks, to be slightly different to those on the Earth, but treating them as if they operated the same, a constant compensation can be achieved.

7. Communication

Transmission of data is required between the Earth and a spacecraft. According to the NASA Thesaurus, "the act of, or methods for, conveying information to, from, or through outer space" is known as space communication. Transmission of information between the Earth and the spacecraft is known as the uplink, and between the spacecraft and the Earth as the downlink. Communication is also required between astronauts and between the teams of scientists and engineers on the Earth. Many missions are now collaborations between different nations and so both the cultural and the language differences of the different parties must be considered to ensure ideas and information between these groups of people are not distorted by translation or cultural assumptions.

Tracking, Telemetry and Command

The information about and the control of a satellite is called the tracking, telemetry and command or TT&C. The tracking element of the TT&C system determines the position of the spacecraft in space. This uses information such as the angle between the spacecraft and the receiver on the ground and the range rate, or the speed or rate at which the range or distance between the receiver and the spacecraft changes. The data from the onboard instruments that monitor the health and position of the spacecraft is encoded and transmitted by the telemetry part of the system. The command element receives, checks and executes commands for the remote control of instruments and the spacecraft. It controls the motion of the spacecraft as well as other subsystems such as the orientation of the solar panels and antennas. The data from experiments or sensors is called the mission data and is also transmitted by the telemetry part of the system, but it is not a part of the system.

Radiowave Communication

TT&C information is usually attached to an electromagnetic wave, such as a radio wave, called a carrier wave. This carrier wave is capable of being modified to transmit data. The simplest way to transmit information between the spacecraft and the Earth is to convert the data into binary digits or bits. This converts all of the data into ones or zeros. These ones and zeros modify the shape of the carrier wave, a process called modulation, and, once transmitted, the information can be decoded or demodulated at the far end. The carrier wave is usually a radio wave as it has a long wavelength that can pass through Earth's atmosphere without too much distortion and it requires less energy to produce than shorter wavelengths.

Disturbances in the carrier wave or in the data itself are called noise. This can interfere with the usefulness of the received data as bits can be lost or misinterpreted and cause errors. Noise can be caused by a variety of factors. Background radio noise or static is radiated naturally from nearly all objects in the universe, including the Sun and the Earth. Man-made interference from mobile phones, power lines, electronic equipment and instruments and also from radio and TV stations can increase the level of noise at the receiver, as can household and industrial appliances and even the local weather at the receiving station. The ratio of the power of the signal received to the power of the noise received is called the signal to noise ratio or SNR.

To minimize the interference between the signals being transmitted to the satellite and those being received from it, different radio frequencies are used. These are usually in different bands or portions of the radio wave spectrum. The International Telecommunication Union (ITU), based in Geneva, Switzerland, allocates the frequency bands that are used by spacecraft. The ITU is an organization within the United Nations system.

Ground Stations

The ground station is the base on the Earth that communicates with the spacecraft. All the information is then passed on to the data users, such as engineers, scientists and astronomers. Different space missions require different networks of communications links, called the communications architecture. For example, a spacecraft that is in a low Earth orbit will often be out of view of the mission's main ground station. Therefore another satellite can be used to relay the data. The relay satellite is often in a geostationary orbit and data can be passed from the ground station to a relay satellite, and then between numerous other relay satellites, before reaching the spacecraft. Passing data between different relay satellites is called a cross-linked network. Another type of architecture is called store and forward, where the spacecraft stores the mission data in its memory and transmits it when it is in view of the ground station. Multiple ground stations can also be used, positioned around the world. These reduce the time a spacecraft will be unable to communicate with at least one location on the Earth.

Some TT&C systems have been standardized, which ensures that different ground stations around the world use the same or similar hardware. The handover from one station to the next is therefore relatively simple and the spacecraft can remain in communication with the ground for most of its orbit around the Earth. The network of ground stations are usually connected via landlines, but some rely on dedicated satellites that forward the data from one station to the next, called relay satellite transponders.

Because different types of spacecraft have specialised needs, there are different standardized TT&C systems. One of these is NASA's Tracking and Data Relay Satellite System (TDRS), which can transmit and receive data and track a spacecraft in a low Earth orbit for at least 85% of its orbit. The TDRS uses three satellites in geosynchronous orbits. Another TT&C system is NASA's Deep Space Network (DSN), which supports interplanetary spacecraft missions. As the Earth rotates a network of three ground stations, located approximately 120° apart around the world, constantly observe the spacecraft. The DSN is described in more detail in Chapter 9 – "Observing Satellites".

Ground stations can be fixed installations on the land, or mobile units on land, sea or air. They are usually located away from heavily populated areas so that the signals that come from the spacecraft are not obscured by radio interference. For most types of spacecraft and mission the ground station requires an antenna, a receiver and a transmitter, which are described in more detail later. They also require data recorders and computers to save and interpret the mission data and the TT&C data.

Spacecraft Systems

The type of communication subsystems on a spacecraft depends on the requirements of the mission, the amount and size of the information that needs to be transferred, the speed of the transfer, known as the transfer rate, and how the data is transferred to the ground station. Other considerations include the maximum Earth to satellite distance, the planned frequency band, launch mass limitations and onboard electrical power limitations. The spacecraft subsystem includes an antenna, transmitter and receiver and the computer support to interpret the received signals into instructions for the control of the spacecraft.

Antennas, Transmitters and Receivers

Antennas, also known as aerials, are used to both direct transmitted radio signals and to collect received signals. They provide the link between a wave travelling though a conductor such as a transmission line, which is known as a guided wave, and what is known as free space, such as the atmosphere or the vacuum of space. Antennas work by the interaction of electric currents. When an electric current flows through an antenna of conductive material, such as metal, it generates an electromagnetic wave. This wave radiates away from the antenna and is therefore broadcast out into the atmosphere or space. The wave and therefore the broadcast is a form of energy and so a constant power supply is required to keep a broadcast continuous. In a similar way, an electromagnetic wave that is travelling though the air or through space will induce an electric current in a conductor, which can be converted back into a signal. An antenna is usually designed to allow the transition from a guided wave to a free-space wave, or vice versa, to be done as efficiently as possible and at the same time to direct the output in a certain pattern. Antennas are required at both the transmitting and receiving ends of the signal and therefore one is required on the spacecraft and one at the ground station. A spacecraft antenna has more demanding requirements than a ground station antenna, as there are weight limitations and also stability problems on the craft and the antenna may also need to be stowed during the launch.

A transmitter generates and amplifies the carrier wave, which is then combined with the data signals. The transmitter then radiates the resulting signal from an antenna. A receiver receives incoming signals from an antenna and converts them into data or instructions. A combination of a transmitter and a receiver into one electronic device is called a transceiver. When the signal arrives at a receiver it is amplified, however this also amplifies any noise in the signal and the amplifier also produces noise. To minimize the presence of noise, low-noise receivers are used, which are cooled to a fraction of the ambient temperature to reduce their noise output.

The gain of an antenna is a measure of the amount the signal is amplified. The angle over which the signal is broadcast, with respect to the direction the antenna is pointing, is called its beamwidth. Most antennas with a wide beamwidth spread the input power, and therefore the signal, over a wide angle and are called low gain antennas. These are used, for example, to cover a large area on the Earth and not just a ground station. Low gain antennas receive more noise than high gain antennas and therefore have a worse SNR than high gain antennas.

High gain antennas produce a narrow beamwidth and so the signal is not spread out and they must therefore be targeted more accurately. If the antenna on the Earth and the one on the spacecraft are not perfectly aligned, the signal may miss and the data could be lost. As the beam from a high gain antenna is very focussed the signal can carry a lot of information. This type of antenna is used to transmit the majority of data collected from a spacecraft back to the Earth. Low gain antennas are always carried on spacecraft as a backup. Most spacecraft have two low gain antennas, one pointing in each direction. This means that it should be able to send and receive signals whichever direction the spacecraft is pointing. Critical problems can then be rectified with signals from the Earth, no matter the orientation of the satellite.

At a geosynchronous altitude, the beamwidth required to cover the visible side of the Earth is 18°. A low gain, and therefore wide beamwidth, antenna, such as a horn antenna or a helical antenna, both of which are shown in Figure 7.1, is often used for this type of transmission. Horn antennas are also used in short-range radar systems on the ground, such as those used to measure the speed of vehicles on a road.

High gain antennas used on spacecraft must use a low transmitting power, as onboard power is limited. The types of high gain antenna most often used on satellites are the parabolic reflector, the lens and the phased array. The satellite television dishes attached to the outside of many houses are parabolic reflectors, as is the Lovell Radio Telescope at Jodrell Bank Observatory,

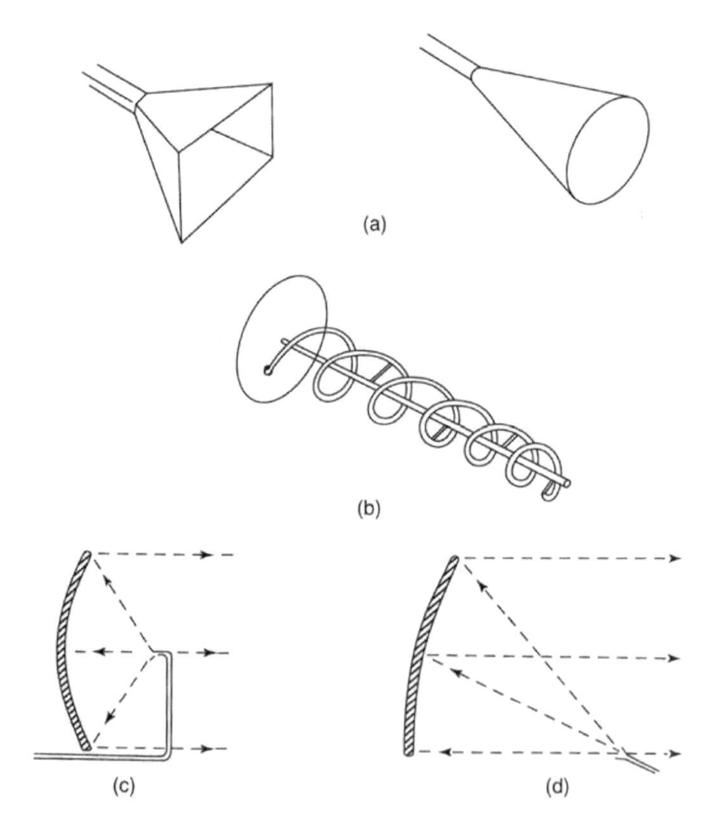

FIGURE 7.1 Antenna Types: (a) Horns (b) Helical (c) Front Feed Parabolic Reflector (d) Offset Feed Parabolic Reflector.

UK. As they have a low mass, low cost and low complexity, as well as having been used for many years and for many missions, parabolic reflectors are often used as the high gain antenna on spacecraft. For example, ESA's *Rosetta* orbiter uses a 2.2 metre diameter parabolic dish. When a reflector is receiving, it reflects and focuses the wave back to a focal point, called the receiver. When transmitting, the signal is fed from a feed located at the focal point and reflected from the dish out into free space.

A lens antenna bends the path of electromagnetic radiation in a similar way to an optical lens that bends light. There are different types of lens antenna, including waveguide, metal plate, dielectric, fresnel and luneberg lenses. Lenses are generally heavier than reflectors, particularly when the diameter exceeds about 0.5 metres. The US military communication satellites, *DSCS III* uses, along with other antennas, waveguide lens antennas for both transmitting and receiving.

Reflectors and lenses are passive and do not radiate or induce a signal themselves, they just focus any signal received. Therefore a feed or a receiver is required at the focus to either transfer the electromagnetic wave to or from free space. The feed can be a low gain antenna, such as a horn antenna.

As spacecraft move in relation to the Earth, the direction of the beam broadcast from the antenna often needs to be changed or it will miss the ground station. Some antennas can be steered physically with motors but the beam direction can also be changed electronically. This method is usually preferred, especially if the beam direction must be changed quickly. This can be done on a reflector by switching to an off-axis feed, where a second feed is offset from the centre. However, this produces high losses in the signal. For a lens, the beam can again be steered by switching between different feed elements but again there are losses in this system. Reflectors and lenses can also be steered by using a phased array feed. A phased array can generate one or more beams simultaneously. By changing the characteristics of each feed electronically, the interference patterns produced either reinforce or suppress the signal, as required by the situation. This interference pattern is similar to the way ripples on a pond either reinforce or cancel each other out.

As well as steering the antenna, narrow beams from high gain lens and reflector antennas can reach multiple ground stations by a process that uses multiple beams. One or more beams can be produced from a switched feed array, which are then hopped to different locations in milliseconds. Instead of one large footprint that can pick up a signal, such as used for a television broadcast, the spacecraft can target different areas with a more concentrated beam and can accommodate two-way communication. As the smaller beams are isolated from each other, the same frequency can be used without them interfering with each other. This method means that high gain antennas can be used to cover a broad area. NASA's *Advanced Communications Technology Satellite (ACTS)* used a reflector and an offset switched feed array to provide multiple hopping beams, and the *Iridium* constellation of communication satellites uses phased arrays to produce multiple beams that cover the Earth.

In radio astronomy, an array of radio telescopes can be linked together electronically to create a synthetic telescope called an interferometer. This creates the effect of one very large telescope that can image astrophysical objects in better detail than any other astronomical technique. If telescopes around the world are used, this system is often called Very Long Baseline Interferometry (VLBI). The data from each telescope is recorded with a highly accurate time stamp, and all the different recordings are later analysed at one control centre. This system can also be used for gathering data from spacecraft. The first time this was implemented was for the *Galileo* spacecraft, when it gathered science data from Jupiter and its moons. Arraying was used to collect more of the weak signal sent from the low gain spacecraft antenna because there were difficulties with the high gain antenna. In this case, the 70 metre diameter dish at Goldstone, California, was electronically linked with an identical antenna at the Canberra Complex in Australia, plus two 34 metre antennas also at Canberra and with the 64 metre Parkes Radio Telescope, also in Australia.

Radio Blackout

At the beginning of the space age, spacecraft regularly lost communication with the Earth, known as radio blackout, as there were not enough ground stations to pick up the signals. With the development of more ground stations, and also the ability to relay data via other satellites, it is possible to communicate with most Earth orbiting satellites most of the time. However, when a spacecraft leaves an Earth orbit, communication becomes more difficult. The first spacecraft to orbit the Moon entered radio blackout when it went around the far side, as the radio waves could not pass through the solid mass of the Moon. *Luna 3*, the Soviet spacecraft that was the first to take photographs of the Moon's far side in 1959, had to wait a day before it could start to transmit the images back towards the Earth. However, it is thought that the signal strength was too low and the transmission was initially unsuccessful. As the probe got nearer to the Earth some photographs were received, although they were quite poor quality. The first manned spacecraft to orbit the Moon was *Apollo 8* on Christmas day, 1968. During their tenth orbit they fired the engines to return to Earth. As this was successful, when they came out of radio blackout, the first words astronaut Jim Lovell spoke to Houston were "Please be informed there is a Santa Claus". Radio blackouts also occur when spacecraft pass behind the Sun or other planets. Some deep space spacecraft shut down their communication systems during their voyage to conserve power, and therefore large parts of their missions are completed without the ability to communicate.

When a storm occurs on the surface of the Sun, the X-rays emitted can cause a disturbance in the Earth's ionosphere, which can cause radio blackouts. The scale and the duration of the storm determines the amount of disturbance to radio communication. The US National Oceanic and Atmospheric Administration (NOAA) classify solar storms as space weather, and their scales are shown in Table 7.1.

Radio blackouts also occur when a spacecraft re-enters the Earth's atmosphere. The friction causes the temperature around the craft to build up heat which strips electrons from the air particles, leaving ionized air around the craft. Communication through the electrically charged ionized air is extremely difficult and communication can be blocked for between 4 and 16 minutes, depending on the shape of the craft and the angle of re-entry. The *Mercury*, *Gemini* and *Apollo* capsules experienced a radio blackout for several minutes during their re-entry phases. Before 1989, the Space

Table 7.1 Space Weather Scale for Radio Blackouts. Courtesy National Oceanic & Atmospheric Administration (NOAA)

Category		Effect		
Scale	Descriptor	Duration of event will influence severity of effects	Physical measure	Average frequency (1 cycle = 11 years)
Radio blackouts			GOES X-ray peak brightness by class and by flux[a]	Number of events when flux level was met (number of storm days)
R 5	**Extreme**	**HF radio**: Complete HF (high frequency[b]) radio blackout on the entire sunlit side of the Earth lasting for a number of hours. This results in no HF radio contact with mariners and en route aviators in this sector **Navigation**: Low-frequency navigation signals used by maritime and general aviation systems experience outages on the sunlit side of the Earth for many hours, causing loss in positioning. Increased satellite navigation errors in positioning for several hours on the sunlit side of Earth, which may spread into the night side	X20 (2×10^{-3})	Less than 1/cycle
R 4	**Severe**	**HF Radio**: HF radio communication blackout on most of the sunlit side of Earth for 1–2 hours. HF radio contact lost during this time **Navigation**: Outages of low-frequency navigation signals cause increased error in positioning for 1–2 hours. Minor disruptions of satellite navigation possible on the sunlit side of Earth	X10 (10^{-3})	8/cycle (8 days/cycle)
R 3	**Strong**	**HF Radio**: Wide area blackout of HF radio communication, loss of radio contact for about an hour on sunlit side of Earth	X1 (10^{-4})	175 days/cycle (140 days/cycle)

		Navigation: Low-frequency navigation signals degraded for about an hour		
R 2	**Moderate**	**HF Radio**: Limited blackout of HF radio communication on sunlit side, loss of radio contact for tens of minutes **Navigation**: Degradation of low-frequency navigation signals for tens of minutes	M5 (5×10^{-5})	350 days/cycle (300 days/cycle)
R 1	**Minor**	**HF Radio**: Weak or minor degradation of HF radio communication on sunlit side, occasional loss of radio contact **Navigation**: Low-frequency navigation signals degraded for brief intervals	M1 (10^{-5})	2,000 days/cycle (950 days/cycle)

a Flux, measured in the 0.1–0.8 nanometre range, in watt per square metre. Based on this measure, but other physical measures are also considered

b Other frequencies may also be affected by these conditions

Shuttle also entered a radio blackout about 30 minutes before touchdown. However, after NASA launched the *Tracking and Data Relay Satellite System (TDRS)* the problem of re-entry blackout was overcome. When the Space Shuttle re-enters it comes belly down and so this part heats up most. At the tail end of the Space Shuttle, where it is cooler, there is much less ionization and this provides a hole through which communications with the TDRS can be maintained. The radio signals therefore go up to a satellite and back down to a ground station, without having to try to pass through the ionized layer around the Space Shuttle.

Blackouts occur whenever a spacecraft enters an atmosphere. The Mars *Pathfinder* mission experienced a 30 second period during which time the communications link to Earth was lost. The *Huygens* probe to Titan did not begin to try to communicate until it had deployed its parachutes and was falling relatively gently through the atmosphere.

Radar

Radar is used to detect and track satellites and other objects in space. It can also be used to gather data about planets and other objects in the solar system. This includes straightforward imaging, three-dimensional modelling and also passive collection of information, such as recording the energy emitted from a planet's surface. The word "radar" is taken from the initial letters of the phrase "RAdio Detection And Ranging".

Radar systems use radio waves to detect both stationary or moving objects or targets. The position of the object, usually in relation to the radar installation, and also the speed and direction of travel can be determined. The radio waves are emitted from a transmitter, often in very short bursts or pulses. If an object is in the path of these pulses it will reflect some of the energy back, just like a cliff reflects back sound waves from a foghorn as an echo. Different shapes of objects and different types of materials reflect the radio waves to a different extent. Sharp internal corners reflect well and so objects that are designed not to be detected, such as military aircraft and ships, have this sort of geometry restricted, which is why stealth planes look strange shapes. Flat reflectors reflect a signal just like a mirror. A receiver is used to detect any echoes that are received during the silence after the pulse. Once the *Cassini* spacecraft entered orbit around Saturn it used radar in three ways to collect different types of data. For imaging, it uses a synthetic aperture radar imager (SAR) that has a resolution of 0.35–1.7 kilometres. An altimeter is used to map surface features with a horizontal resolution of 24–27 kilometres and vertical resolution of 90–150 metres, and a radiometer maps the infrared radiation emitted by the planet with a 7–310 kilometres resolution.

Different types of radar systems are used for different purposes. For example, a phased array radar system is used to track multiple satellites at the same time. This system consists of a group of antennas where each outgoing signal is at a different phase. Like the phased array radio, the interference patterns produced by the phased array radar either reinforce or suppress each other as required by the situation. By manipulation of the signals, the signal can be broadcast or received at an angle to the plane of the array, and it can therefore be steered, without actually moving the antennas. As the radar has no moving parts, it can switch to scan different areas of the sky very quickly. The antennas for radar phased array systems do not have to be spread out like the radio telescope interferometers, but can be located next to each other.

Parcels to Space

Some things are required in space that cannot be sent electronically. The International Space Station is the main object that requires supplies from the Earth, including fuel for station-keeping and food and water for the astronauts onboard. The *Progress* resupply vehicle, an automated, unpiloted version of the *Soyuz* spacecraft, usually services the ISS. It is launched from the Baikonur Cosmodrome in Kazakhstan and normally takes two days to reach the Space Station. Both the rendezvous and docking are automated, although once the spacecraft is within 150 metres of the Station both the crew and the Russian Mission Control Centre monitor its approach and docking. *Progress* can carry up to 1,700 kilograms of supplies to the Space Station in a pressurized volume of about six cubic metres. After the crew has removed the cargo, they refill it with rubbish, unneeded equipment and wastewater, which burn up with the spacecraft when it re-enters the Earth's atmosphere.

Parcels from Space

Some things gathered from space cannot be returned electronically and so must be physically sent to the Earth. Between August 1960 and May 1972, the *Corona* satellites were used by the USA as an Earth observation or reconnaissance system. The program and images were declassified in 1995. The satellites carried cameras and were used to photograph the activities of other nations, mainly the Soviet Union, during the Cold War. The film canisters were returned to Earth to be processed and viewed. As the photographic coverage was worldwide, the images were also used to produce maps and charts for the Department of Defence and other US government mapping programs. The film canisters were returned to Earth in capsules known as buckets. Specially equipped aircraft recovered the buckets as they descended by parachute, but the buckets were also designed to float so they could be recovered from the ocean. All the film was black and white, although a small sample of infrared and colour film was carried on some missions as experiments. Since declassification, the 800,000 images have been used to aid environmental studies, as they are the only images from space available for that time period. They are used to analyze changes that have occurred over time in weather patterns, ocean temperatures, landmasses and other features.

As mentioned in Chapter 3 – "Space Missions", in January 2006, NASA's *Stardust* return capsule successfully brought back samples from comet Wild 2 and landed by parachute in the US Air Force's Utah Test and Training Range. Also mentioned previously, there was a less successful sample return mission in September 2004. NASA's *Genesis* mission collected particles of solar wind in a capsule and during the parachute descent a helicopter was meant to recover it from mid air. The parachute did not open and the capsule hit the ground at over 300 kilometres per hour. Fortunately, some useful information has been extracted from the wreckage.

Human Communications

Most space projects are now multinational collaborations and citizens of over 30 countries have flown in space. Currently, every astronaut, cosmonaut and even taikonaut is required to have a good knowledge of English. Astronauts who visit the International Space Station must also learn Russian as many systems, including the electrical power supply, communications system and the current emergency escape pod, a *Soyuz* space vehicle, are all Russian built and all the procedures and instructions are written in Russian. NASA and ESA astronauts spend time in Russia for advanced training prior to launch. This is because knowledge of not only the language but also the culture is very important and helps crews of mixed nationalities understand each other better. The working language on the ISS is English, although the Russian flight controllers, the equivalent of Mission Control in the USA, communicate in their native language. As all flight crew must understand commands being sent to their spacecraft, a good working knowledge of Russian is essential.

The language differences are the most obvious problem of working in an international team. However, cultural differences can also cause problems or misunderstandings. For example, a shake of the head in one culture means "no", whereas in another it can mean "maybe". The British think of a switch as something that is "switched" up and down for off and on, whereas Americans turn their switches clockwise and anticlockwise. The colour red in Western terms generally indicates a warning or danger, whereas in the Far East red symbolises good luck. There are even differences in how the astronauts eat. Americans are seen to have lunch simply to refuel their bodies, whereas continental Europeans see it as a time for breaking off from work and time for a social interval.

The attitude towards alcohol also appears to vary between nations. The US policy is not to allow alcohol aboard flights, whereas aboard the *Mir* Space Station, cognac was apparently available, although maybe not with the full approval of the authorities. Alcohol has been taken into space on NASA missions, but only Buzz Aldrin has consumed some on a mission. He silently took Holy Communion with wafers and wine in the *Eagle* landing craft after he had landed on the Moon. In his book *A Man on the Moon, the Voyages of the Apollo Astronauts* Andrew Chaikin describes Aldrin pouring the wine into a chalice:

> Released in the gentle gravity the wine poured slowly and curled gracefully against the side of the cup.

Other alcohol that has been taken aboard NASA missions but not drunk, includes three tiny bottles of brandy that were on the *Apollo 8* mission as a Christmas treat. The crew did not drink the bottles, as if they made a single mistake on the return journey, they thought the alcohol would get the blame. The Frenchman Lieutenant Colonel Patrick Baudry, who flew as a payload specialist on STS-51-G aboard *Discovery* in June 1985, took some wine into space. He is a wine connoisseur. His crewmate John O. Creighton, who was interviewed for NASA's oral history project, said:

> Patrick carried four very small vials of wine up there, and they were sealed and we couldn't get our hands on them, but anyway, he took one of them back and presented them to the Bordeaux Wine Society.

He added:

> But we didn't have access to it; it was down below the floorboards someplace.

Aboard the *Mir* space station, the attitude towards alcohol appeared a little more relaxed. Dr. Norman Thagard, a US astronaut who became a cosmonaut-researcher on *Mir* on mission *Mir 18* in 1995, was interviewed as part of NASA's Shuttle-Mir Oral History Program in September 1998. He says:

> A couple days before launch, we were decanting this cognac into these plastic bottles. We wrapped them real tight with tape, after screwing the cap on as tightly as we could, and then wrote what in English would look like COK, but is SOK in Russian, 'sok' is the word for juice. So we labelled them all 'juice', and they carried them out and put them in the Soyuz. So we launched with quite a lot of cognac on the Soyuz.

The mission lasted 115 days, and he adds:

> Typically on a Friday night, Gennady [Gennady Strekalov, the flight engineer onboard] would come into the base block there at

> the dinner table and have a little smile on his face, and he'd look at me, and he'd say in English 'It's time to have a drink'.

Dr. Thagard recalls:

> I had my fifty-second birthday when the Shuttle was there, the Russians gave me a bottle of really good Russian cognac.

However, no one seemed to drink to excess.

> I never saw anybody drunk. In fact, I never saw anybody drink more than what I would consider to be one shot. Because how would we do it, we'd take the cap off one of these plastic bottles, and even with alcohol there's enough surface tension that it tends to adhere to the side, it just doesn't come gushing out, and as long as you don't agitate the bottle, its contents tend to stay in there. We all had our straws, and we would take straws and we would stick them down into one of the bubbles and just take a sip, and then you would gently push the bottle and it would float over to the next person, and they'd put their straw in.

Radio and Video Links

Astronauts aboard the ISS can send emails and also make telephone calls to friends and family anytime they like. This is done through an Internet Protocol (IP) phone, routed through computers and Ku-band satellites. Astronauts also have a weekly videoconference with their families. The families are in their own homes, which gives the astronauts a feeling of being at home with their family. This is very important psychologically and helps the astronauts feel connected to the ground and less isolated, especially as some astronauts comment that they feel like they are circling the globe in a tin can.

Amateur Radio

The *Mir* space station was equipped with a small and simple low powered VHF Amateur or Ham radio station. It was mainly used to provide a connection for the cosmonauts onboard to family and some new friends back on Earth. However, it also was a "back-up" for radio communications to Earth and was used for this purpose more than once.

The ISS currently has two sets of Ham radio equipment on board, with a third planned for the European Columbus Module.

The first set is located in the Functional Cargo Block (FGB), Zarya. The equipment consists of a handheld 5 watt VHF FM transceiver, headphones with a microphone attached and an outboard antenna system. This is the same type of equipment in use around the world by many amateur radio enthusiasts. Packet radio is also available. This is similar to sending text messages on a mobile or cell phone, but instead of using a phone, the operator uses a computer, modem and the radio transmitter. The second set of Ham radio equipment is in the Zvezda Service Module, the crew's living quarters. The equipment there comprises of two multiband transceivers. Four antenna systems were deployed during three spacewalks or extra-vehicular activities (EVAs) and clamped onto handrails on the outside of the Service Module. They provide operation on the HF (20, 15 and 10 metres), VHF (2 metres), UHF (70 centimetres) and the microwave bands (L and S band). These have a dual use as they also permit the EVA video signals to be received.

You no longer have to be the President of the USA or a VIP to communicate with the astronauts aboard the ISS. Anyone with a scanner can listen to the communications that take place by radio between the Earth and the ISS as it passes overhead, and if you have a transmitter, and an amateur radio licence, you can take part in the conversations. The Amateur Radio on the International Space Station, or ARISS, program offers the opportunity for students to join in and talk directly with crewmembers on the ISS. This is sponsored and supported by Amateur Radio organizations and space agencies in the USA, Russia, Canada, Japan and Europe.

The ISS is only overhead for about ten minutes at a time in any given area, so you have to be prepared. Information such as when the Space Station will be over your location, what frequency the astronauts transmit on and the crew's schedule and their call signs, can be found on NASA's and ARISS's web sites. After you have made contact, you can request a postcard-type document, called a QSL card. This proves you have spoken to an astronaut. Further information on this is given on ARISS's web site. There are some northern and southern areas of the world where you cannot get direct reception from the ISS, for example, in some areas of Australia and South Africa. However, a program called Tele-bridge can be used, which is a phone bridge set up to communicate from a telephone to short-wave radio.

8. Humans in Space

Outside of the Earth's protective atmosphere, human survival is only possible with the aid of life support systems. Many of the conditions in space can be reproduced on the Earth, such as the high temperatures and low pressure, but the reduced gravity cannot be simulated. When Laika the dog was launched into space by the Soviet Union in *Sputnik 2* in 1957, many scientists did not believe anything would survive in reduced gravity. Even up to the first human space flight in 1961, when Yuri Gagarin circled the globe in *Vostok 1*, there was still doubt whether humans could adapt to live and work in space. This is why Gagarin's space flight was limited to just 108 minutes and a single orbit.

Launch and Re-entry

The human body can only tolerate a certain amount of physical stress before permanent damage occurs. Manned spacecraft must be designed with these limitations in mind. The physical stresses of most concern during the launch are acceleration, noise and vibration.

Acceleration

The force felt as a result of acceleration cannot be blocked, unlike sound that can be diminished by wearing ear defenders. Therefore an astronaut must endure the force from the acceleration at launch and the maximum acceleration of the rocket must be restricted to levels the human body can tolerate.

Acceleration is often measured in *g* force. The Earth's gravity has a force of 1 *g*, and the effect of this force on your body results in your feeling weight. If you are in a lift and accelerate upwards you will feel heavier than normal. If you accelerate downwards you will feel lighter than normal. If you accelerate downwards at a rate of 1*g*, you will feel weightless. If you are travelling horizontally and accelerate you will feel heavier than normal, if you decelerate you will feel lighter. The tolerance an individual has to acceleration forces is dependent on many factors, including the size and the direction of the force in relation to the position of the body. An acceleration of 2*g* just makes you feel heavy, whereas 100*g* will probably kill you. Tolerance is lowest when the force acts from your feet towards your head and greatest when laying on your back and the force acts from your back to your chest. At increased levels of *g*, two main effects occur. First, the heart must work harder to pump the blood up to the head, as it has more force to overcome, secondly, blood pools in the legs as the extra force of the acceleration also impairs the normal return through the veins. Therefore the amount of blood flowing through the heart with each beat is reduced, which further reduces the blood pressure and the amount of blood reaching the brain. When too little blood reaches the retina of the eye, the peripheral vision deteriorates until the field of view is reduced to tunnel vision and colour vision is lost. This is also known as grey-out. If the force is sustained beyond this point, vision then becomes hazy, until nothing can be seen. This is known as blackout. Hearing is not impaired at this stage, and you can think normally, although with sustained or increased force unconsciousness will occur, known as *g*-induced loss of consciousness or G-LOC. If the force is in the opposite direction and blood pools in the head, red-out can occur, when only red can be seen. When normal blood flow is returned to the eyes, these visual symptoms disappear immediately.

There are other effects on the body due to acceleration forces. The liver sinks into the abdomen and the heart descends in the chest. This pushes the diaphragm downwards, which makes breathing more difficult. Movement of the limbs becomes difficult and after about 4*g* it is not possible to move the head. At 8*g* it is almost impossible to move the arms, and at 25*g* hand movement is only just possible, as can be seen in Figure 8.1.

The ability to withstand g force is also reduced when the oxygen levels in the air supply are low and the environment is at high temperatures. The amount of training the individual has had and the amount of time they have spent in weightlessness immediately before the onset of the force also affects the amount that can be tolerated. In the

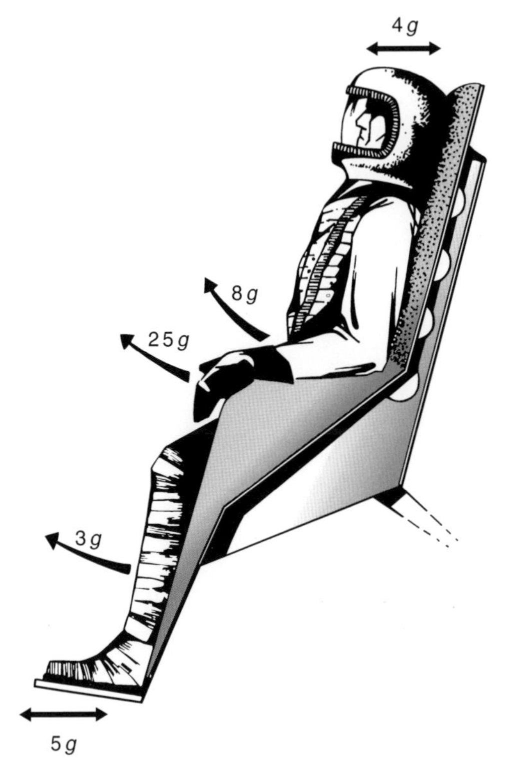

FIGURE 8.1 Acceleration at Which it is Just Possible to Move Various Parts of the Body.

1940s and 1950s Dr. Stapp of the US Air Force pioneered experiments on the effects of acceleration on the human body and was himself the guinea pig in many of the experiments. He demonstrated that a human could, with adequate harness, withstand at least 45*g*. This is the highest known g forcc voluntarily encountered by a human.

During launch and landing, astronauts lay on their backs with their legs bent. This reduces the height the heart must pump the blood to the brain as they are almost on the same level. Further, by raising the legs and bringing them closer to the heart, the blood does not pool in the legs. In the Space Shuttle, the astronauts feel between 2 and 3*g* during launch and re-entry. There are no special seating arrangements on the Space Shuttle, so any astronaut can use any seat. However, aboard the *Soyuz* craft, the crew feel gravitational forces of up to 4*g* or even higher. Each crewmember aboard a *Soyuz* has a custom-built seat or couch liner that distributes the shock of impact and prevents injuries during the landing.

Figures 8.2a and b show one of these seats. In the past, some cosmonauts have suffered minor injuries such as bumps and bruises during landing aboard a *Soyuz*.

On the Russian *Soyuz* craft, the sensation of increased gravitational force on the body during landing includes laboured breathing and speech and an increase in weight. Anousheh Ansari, the first female private space tourist, described it:

> like an elephant was sitting on my chest.

Many astronauts feel a lump in their throat and to prevent further discomfort, they try not to swallow or talk. To help counteract the effects, an anti-g straining manoeuvre can be performed. This involves clenching muscles in the lower body, whilst relaxing the upper body and breathing steadily.

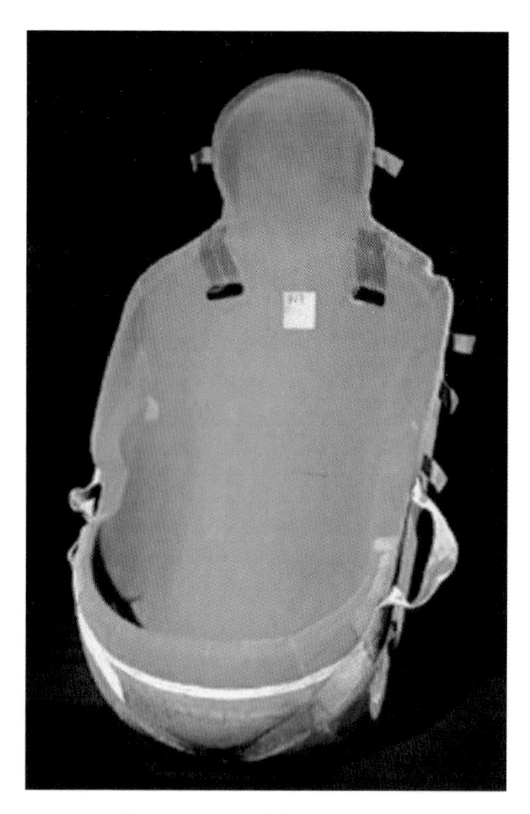

FIGURE 8.2a Empty Couch Liner Used by NASA Astronaut, Dr Norman E. Thagard Onboard a Soyuz in 1995.
Photo by Mark Avino, National Air and Space Museum, Smithsonian Institution (SI 97–15147).

FIGURE 8.2b Couch Liner with Model, Used by NASA Astronaut, Dr Norman E. Thagard Onboard a Soyuz in 1995.
Photo by Mark Avino, National Air and Space Museum, Smithsonian Institution (SI 97–15147)

Vibration and Sound

During the launch, the rocket engine causes vibrations throughout the whole craft. Although the duration of these vibrations is usually quite short, they can be intense. Instruments and equipment, including all their electrical and mechanical connections, must be rugged enough to withstand these vibrations. The effect on the human body, however, must be considered when designing the spacecraft. Vibrations can cause body parts to move or even become displaced, which can be uncomfortable or even painful. The muscles try to control this movement and they rapidly tire, causing further discomfort. Vibration in the low-frequency range can also cause blurred vision, shortness of breath, motion sickness, and chest or abdominal pain and moderate vibration slightly increases the metabolic rate. Some frequencies have been shown to affect the liver, spleen, and stomach, and others may result in mouth, throat, bladder or rectal pain.

Engine noise is the most obvious cause of sound inside a spacecraft during launch. However, until the rocket reaches supersonic speeds, the buffeting of the rocket as it pushes through the air also produces noise. The level of noise reaching the astronauts must be reduced to prevent either permanent or temporary deafness. However, it is not just the level of noise that can affect the human body. Low-frequency sounds can be felt through the surface of the body. This can cause a similar effect to vibration and can cause the internal organs to vibrate and can disturb vision. Extremely low frequencies can mimic brain rhythms and cause severe disturbances in mental functions.

Environmental Control and Life Support

Everything an astronaut requires in space has to be brought up from the Earth. The early space missions had life support systems designed to be used once only. Oxygen for breathing was provided from high-pressure storage tanks. The carbon dioxide produced by the astronauts was removed from the air by a chemical reaction with lithium hydroxide in replaceable canisters. The water for the *Mercury* and *Gemini* missions was stored in tanks, while the electricity producing fuel cells on the *Apollo* spacecraft provided water as a by-product. Urine and other waste water were collected and either stored or vented overboard. It is expensive to transport anything into space, and for missions outside of the Earth's orbit, it may not even be possible to resupply a spacecraft from the Earth, due to the vast distances involved. Therefore water and oxygen will need to be recycled whenever possible, by the onboard environmental control and life support systems. These systems are already used to control the air quality, the air pressure, humidity, temperature and fire suppression.

Air Quality

On the International Space Station the oxygen and nitrogen levels, as well as the pressure, are kept the same as they are at sea level on the Earth, which is about 21% oxygen and 78% nitrogen at 1.01 bar. Oxygen is supplied primarily by the Elektron electrolysis system, which separates water molecules into hydrogen and oxygen. Although this is more efficient than supplying oxygen in tanks, it is not recycling and therefore not sustainable for long duration missions that could not rely on refills of water coming from the Earth. Additional oxygen, nitrogen and pressurization services are available on the ISS from refillable tanks.

After breathing in oxygen, we breathe out carbon dioxide. On the ISS this is removed from the air by a machine on the Zvezda Service Module that uses a solid, porous and absorbent material called zeolite, which acts as a molecular sieve. We also produce small amounts of other gases, including methane in the intestines and ammonia by the breakdown of urea in sweat. We also emit acetone, methyl alcohol and carbon monoxide in our urine and breath. To prevent these gases from accumulating in the spacecraft, activated charcoal filters are usually used. These are similar to the chemicals used in water filters you can buy for your home. As the charcoal traps the gases, they will eventually fill up and must therefore be replaced regularly.

Pressure

Although there are towns and villages around the world that are higher than 5,000 metres above sea level, most people live in the lower atmosphere, at an altitude of less than 1,500 metres. As the altitude increases, the air pressure falls and the air thins out. There are the same proportions of nitrogen, oxygen and other gases, but there are fewer molecules. Therefore, as the height increases, the amount of oxygen taken into the lungs with each breath decreases. High altitude is usually defined as 1,500–3,500 metres, very high altitude is between 3,500 and 5,500 metres and extreme altitude is above 5,500 metres. At increased altitudes, the body compensates for the lack of oxygen molecules by breathing faster or deeper or both, known as hyperventilation. Other effects include shortness of breath during exertion, a change in breathing pattern at night, frequently waking up during the night and increased urination. Although the breathing rate is increased, the normal level of oxygen in the blood cannot be attained at high altitudes. Lack of oxygen in the body and brain is called hypoxia. The symptoms, such as reduced concentration, clumsiness, faulty judgement, moodiness, headache, deterioration of vision and a high pulse rate, may not be noticeable at first, as a feeling of euphoria may mask any problems. Finally loss of consciousness will occur. The time in which someone who is suddenly deprived of an adequate oxygen supply can still perform useful tasks before hypoxia sets in is called the time of useful consciousness. This gets shorter with higher altitudes. At about 6,500 metres, which is about 300 metres higher than highest mountain peak in North America, Mount McKinley in Alaska, it is about five minutes, whereas at 12,000 metres there is only about 12 seconds of moderate activity before the person loses consciousness. As passenger aeroplanes travel at heights of about 9,000 metres, the cabin is pressurized or pumped to that found at an altitude of about 1,500 metres. This way the passengers and crew are not subjected to significantly reduced oxygen levels.

The pressure at sea level is about 1.01 bar, at 1,500 metres it is 0.84 bar and at 10,000 metres it is 0.26 bar. If the pressure is as low as 0.11 bar, which it is at about 15,000 metres above sea level, the body cannot take in any oxygen, as the alveoli in the lungs become saturated with water vapour and carbon dioxide. Even though there is still oxygen in the air at this altitude, it cannot enter the lungs and any attempt to expand the chest just causes the body to pump more carbon dioxide and water vapour into the alveoli. This leads to suffocation or drowning in your own water vapour.

If an unprotected body were subjected to 0.06 bar, equivalent to an altitude of about 19,000 metres, bubbles would form in body fluids, known as ebullism. This would first occur in the mucous linings of the mouth and the eyes. This is what happens when a fizzy drink bottle is opened and the gas fizzes out. In 1965, NASA accidentally subjected someone to a near vacuum. A man in a leaking spacesuit was taken down to about 0.07 bar in a vacuum chamber. He remained conscious for about 14 seconds, about the time it takes for oxygen deprived blood to go from the lungs to the brain. The chamber pressure was increased within 15 seconds and the subject regained consciousness at about 0.57 bar. He later said that he could feel and hear the air leaking out, and his last conscious memory was of the water on his tongue beginning to bubble. The same thing can happen to scuba divers who ascend from deep water too quickly. When they are deep underwater they are subjected to large pressures, and gases, such as nitrogen, are absorbed into the blood. When the pressure is reduced, the gasses may form bubbles in the blood stream. This is known as decompression sickness or the bends.

If an unprotected astronaut were exposed to space, as long as they did not try to hold their breath, they could probably survive for about 30 seconds without permanent injury. If they held their breath, they would probably damage their lungs as the gases try to expand. However, they would not explode and their blood would not boil because the skin and circulatory system would prevent this. They would not freeze, as, although space is cold, a warm body does not radiate heat away very quickly. They would not lose consciousness until the oxygen in the body was used up. They may get sunburnt by the unfiltered UV radiation from the Sun, and suffer other problems like the bends and swelling of the skin and tissue. However, if they had not been recovered within a couple of minutes, they would probably die.

Temperature

One of the hottest places people live on the Earth is Dallol in Ethiopia. It has an average temperature of about 34°C. By comparison London in the UK, has an average temperature of about 12°C. One of the coldest inhabited places is Oymakon in eastern Siberia, Russia, where it has reached a minimum of nearly –68°C. On the Earth's surface the temperature can range from –88°C to 58°C. On the Moon it ranges from –233°C to 123°C.

The human body has a temperature of about 37°C, although this varies in different people between about 36.5°C and 37.2°C. If the body's temperature gets hotter than this you begin to feel ill and if it rises by about 7°C you could even die. Your body can stand to lose more heat than it can gain, but if it drops below about 26°C, hypothermia will set in and you will probably die. There have been people who have survived extreme cold for hours or even days, including the Japanese man Mitsutaka Uchikoshi, who was revived after being unconscious for 24 days on Mount Rokko in western Japan in October 2006. Doctors are not sure how these people survived but using extreme care when warming them up, most have made a full recovery.

The body's temperature is controlled by the hypothalamus, a part of the brain that acts like a thermostat. Temperature is regulated by balancing the heat produced within the body with the heat lost through evaporation of perspiration, radiation, convection and conduction. If the environment is warm, heat must be lost from the body. This is done by increasing the blood flow to the surface of the body, which causes the flushed look, where the heat can be radiated out. If the environmental temperature is above about 34°C, or if the body's temperature has been increased by manual work, heat is lost through evaporation of sweat. However, if the humidity is high, evaporation of sweat is slower and perspiration is not an effective method of cooling. When the body temperature gets too high it is called hyperthermia.

In a cold environment, the blood vessels near the surface of the body constrict, restricting the blood flow to these areas and therefore limiting the heat lost by radiation. A wave of muscle contractions, better known as shivering, increases the body's metabolism and therefore increases the internal temperature. Goose bumps are caused by the hairs on the skin raising, although this is not particularly effective in humans. For animals with feathers or fur this increases the thickness of their insulation. Below about 4°C, without clothes, a person cannot increase the metabolic rate sufficiently to replace the heat lost. Hypothermia, will set in and can cause death. But it is not just the external temperature that effects heat loss. Air movement or wind chill or contact with water or metal can rapidly increase heat loss and hypothermia can result more quickly under these conditions.

As most electronic and mechanical equipment also works best within certain temperature ranges, the thermal control of a spacecraft is important for both manned and unmanned spaceflight. The temperature and humidity is controlled on the ISS by the air being pulled through filters that cool and dehumidify it, before it is recirculated back through the station. This also ensures the air moves within the station, as there are no natural convection currents. Convection, caused by warm air rising depends on gravity. Although the warm air still expands in freefall conditions, and it has a lower density, it is no longer lighter than cold air as they are both "weightless". Without air movement astronauts would get surrounded with bubbles of carbon dioxide from the air they have breathed out, and they could suffocate.

Fire Suppression

A fire aboard a spacecraft acts differently than one on the surface of the Earth. On the Earth the gravity allows the hot gas to rise and the flame becomes long and pointed. In freefall, the hot gasses do not rise, and instead of a tall yellow flame on a candle, for example, a smaller, blue flame centred on the wick, will be seen, as in Figure 8.3.

Although in space there are no convection currents to help the fire spread, there are numerous ventilation fans to ensure fresh air is moved around the craft, and this air movement will replenish the air supply to the fire and also enable the fire to spread in any direction, rather than just up. The flames in space can also produce different amounts of soot, smoke or harmful gases compared to when in Earth's gravity. To detect a fire in the air supply of the ISS, smoke detectors are placed in the ventilation system. As smoke will not rise in space, there is no point putting one on the ceiling. If a fire is detected, alarms are sounded and the astronauts turn off the ventilation system. This removes the air currents and decreases the amount of oxygen available to the flames. It also stops the smoke travelling throughout the

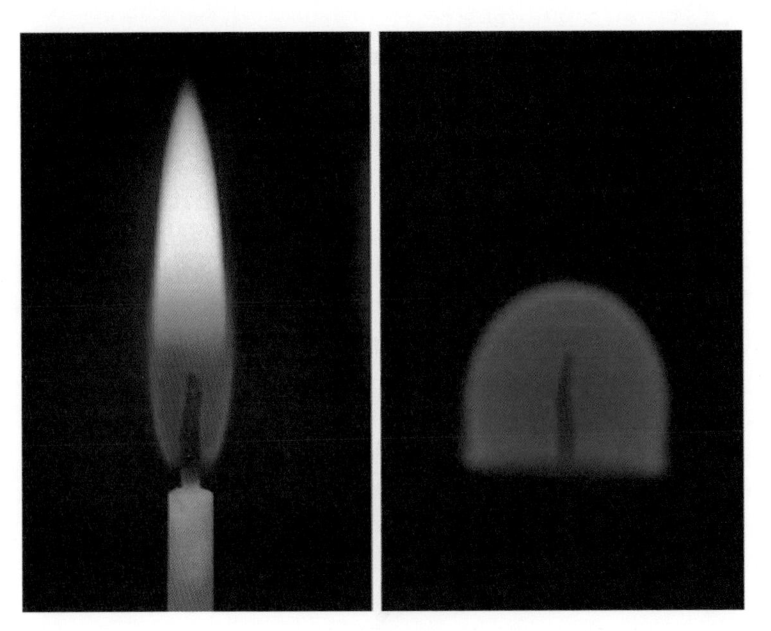

FIGURE 8.3 These Two Flames are Burning on the Same Type of Candle and are Shown at the Same Scale. The Image on the Left is Burning Under Normal Gravity, the Image on the Right is Under Microgravity.
Image courtesy NASA

station. Electrical power is then shut off from the area where the fire is and fire extinguishers are used. There are two types of fire extinguishers aboard, water-based foam units and carbon dioxide units. The water-based foam units work best on fires of paper, wood and plastic and the carbon dioxide units are better on electrical fires.

Water

An astronaut needs about 2.7 litres of water every day, which is gained through their food and drink. Most of this water leaves the body again, either as a liquid, as urine or sweat, or as a vapour through breathing and through pores in the skin. They also use about four litres a day for personal hygiene, which is much less than they would use on the Earth, where a shower uses about 50 litres and flushing the toilet uses about six litres. If the water onboard a spacecraft was not recycled, there would either need to be frequent resupply missions, or the length of the astronaut's stay would be limited. When the ISS is completed, most of the water will be recycled. Some water will be lost as brine in the water recycling systems and some will be lost as humidity in the air when the air lock is opened. The carbon dioxide removal system also removes some water out of the air. Water is replenished on the ISS, either from the Russian *Progress* rockets, which can be fitted to carry large containers of water, or from the Space Shuttle. Like the *Apollo* spacecraft, the Shuttle produces electricity using fuel cells, and therefore also provides water, which it can transfer to the ISS when required. Although the fuel for the fuel cells must also be lifted up from the Earth, the water produced is a useful by-product of the electricity generating process.

If the water vapour in the air were not removed the astronauts would soon feel as though they were in a sauna and would have difficulty breathing. On the ISS the water in the air is recovered by condensation. The warm, humid air is blown across a cold surface and tiny water droplets form. However, the droplets do not run down the surface as they would on Earth, so they need to be collected another way. The surface is spun so the droplets move to the outside, where they can be sucked away. Another method uses a surface coated with a substance that attracts water and makes the water stick to it. Tiny holes in the surface, connected to suction tubes called slurpers, then suck the water away, which is then broken down to into its components of hydrogen and oxygen. The hydrogen is vented overboard, and the oxygen is put back into the cabin's atmosphere. At the moment, all waste from the toilet is disposed of. However, in the future, urine will be collected from the toilet and the water recovered using a low-pressure vacuum distillation process. The entire process will occur within a rotating distillation assembly that, like the spinning condensation surface, will compensate for the absence of gravity and help to separate liquids and gases in space.

Radiation

Radiation is energy that is transmitted in the form of waves or subatomic particles, such as electrons, protons and ions. On the Earth, electromagnetic radiation is the most common form of radiation. This includes X-rays, gamma rays, visible light, infrared and microwaves. X-rays and gamma rays have enough energy to knock the electrons out of any atom that they strike, which means they can ionize atoms. This form of radiation is called ionizing radiation. Radiation that cannot ionize other atoms is called non-ionizing radiation. As ionizing radiation causes damage to atoms, humans and other living creatures must be protected from it. It can also damage sensitive instruments such as electronic components in computers, and precautions must be taken to protect critical systems.

Although X-rays and gamma rays are present in space, the majority of space radiation is not in the form of electromagnetic waves. It is composed mainly of subatomic particles. There are three main sources of space radiation. These are particles that are trapped in the Earth's magnetic field, making up the Van Allen belts, as described in Chapter 1 – "Introduction", solar particle events and galactic cosmic rays.

Most of the particles in the Van Allen belts originate from the solar wind. The lower belt consists mainly of high-energy protons, the outer belt contains mainly high-energy electrons. Most manned space missions have so far kept below the lower Van Allen belt, and as such have been protected from the radiation within it.

They are also protected from the solar wind by the Earth's magnetic field. However, some low-inclination flights pass through the South Atlantic Anomaly, an area where the inner belt is only about 200 kilometres above the surface of the Earth. Spacecraft on an orbit that passes through this area are subject to larger amounts of radiation at these times. Also, high-inclination orbits that go over the Earth's polar regions also receive larger amounts of radiation, as the Earth's magnetic field does not protect these areas.

Solar particle events occur when activity on the surface of the Sun, such as solar flares and coronal mass ejections or CMEs, causes electrons, protons and ions to be ejected into space. Electromagnetic radiation in the form of X-rays and gamma rays are usually also released by these events, however, these forms of radiation do not usually have sufficient energy to cause harm to astronauts in low Earth orbit. Most solar particle events do not pose a particular radiation hazard, as they are either too small or their path misses the Earth and orbiting spacecraft. Most particles are deflected by the Earth's magnetic field. However, occasionally, a large solar flare or CME does erupt and so the Sun is constantly monitored and when a threat occurs precautions are taken to protect astronauts and sensitive equipment.

Galactic cosmic rays or GCRs, are the major source of space radiation. The Sun's magnetic field deflects these particles, so when the field is at its weakest, during solar minimum, the GCRs are at their highest intensity. The solar particle events are at their lowest intensity during this time. Cosmic rays are difficult to shield against and are therefore often more hazardous than occasional solar particle events.

The immediate effects of ionizing radiation on the human body include skin-reddening, vomiting or nausea and dehydration. If the dosage is not too high, the astronaut can recover. However, if the dosage is higher, or sustained over a long period, further damage may occur. The atoms in the cells, including the genetic material DNA, could be damaged. This may cause the cell to no longer carry out repairs and reproduction correctly, and may lead to mutations that can cause cataracts, tumours, cancer, genetic defects in offspring, or death.

The main countermeasure against radiation is to limit the amount of exposure. This usually means limiting an astronaut's time in space. However, with long duration missions, such as a manned trip to Mars, other precautions must be taken, especially as the Earth's magnetic field will not be available to provide protection. The fabric of a spacecraft blocks some of the radiation but it does not completely shield the astronauts. The best material to block high-energy radiation is hydrogen, but a shield out of pure hydrogen is impractical with current technology. Materials with a high hydrogen content, like polyethylene, the material supermarket bags are made from, could be used. Water is also good, but it is heavy and therefore expensive to launch. To block radiation completely, hydrogen-rich shields would need to be a couple of metres thick, although a shield of only about seven centimetres thick can block 30–35% of the radiation. Space walks or EVAs are not planned if a solar particle event is expected, as an astronaut's spacesuit will not provide sufficient protection. Each astronaut aboard the ISS wears a physical dosimeter to measure the radiation they receive during their flight and there are also sensors throughout the ISS itself that measure radiation exposure.

When astronauts are aboard the ISS, NASA flight surgeons monitor the solar weather, using data supplied by the US National Oceanic and Atmospheric Administration (NOAA). If there is a radiation environment alert the flight rules read:

> Inform Crew. Recommend Remaining In Higher Shielded Areas And Avoiding Lower Shielded Areas During Intervals Of High Risk Orbital Alignments.

This basically means that the astronauts should go to an area of the station that has a lot of water or polyurethane foam, or both. These areas include the aft end of the Zvezda service module and the Unity Node, where the Russian and US water tanks and bags are stored and the US sleep station in the Destiny laboratory, which is made of a polyurethane material and also has water around it.

Radiation is one of the major concerns for long duration manned space flights and methods of minimizing the risk are being investigated. Researchers are working on protective substances that may be taken by mouth or injection prior to radiation exposure to limit the damaging effects. It has been found that antioxidants such as vitamins C and A reduce the damage caused by radiation and work is also being conducted into ways to help the body after the damage has occurred, for example, damaged cells could be instructed to destroy themselves.

The *Apollo* astronauts reported seeing strange flashes of light inside their eyeballs, even with their eyes closed. Astronauts aboard *Skylab*, the Shuttle, *Mir*, and the ISS have also reported seeing these flashes. They are caused by radiation from space hitting their eyes like bullets. When one of these "bullets" strikes the retina, a signal is triggered that the brain interprets as a flash of light. This is not good for the eyes, and years after returning to Earth, many astronauts have developed cataracts, where the lens of the eye is clouded. By 2001, 39 former astronauts had suffered some form of cataracts after flying in space. Of those 39 astronauts, 36 had flown on high radiation missions, such as those that went outside of the Earth's protective magnetic field and Van Allen belts on their way to the Moon. Some cataracts appeared four or five years after the mission, but others took ten or more years to manifest. The link between radiation and cataracts is still not fully understood.

On the ISS there are some windows that do not filter out all of the UV radiation, as they are required for special photographic use. When near these windows, the crew must be particularly careful to protect their eyes and skin. Special sunglasses have been developed to protect astronauts' eyes. These sunglasses have a thin layer of gold across the lens, which offers better UV and infrared protection. These glasses provide 100% UV protection and permit only 5.5% of the light and just 8% of blue light through. Blue light has also been shown to damage the eye. The frame design also provides protection as the strongly curved form reduces light penetration from around the edges of the glasses.

Meteoroid Strike and Space Debris

In space there are many small bits of ice, rock and metal, called meteoroids, travelling at average speeds of about 20 kilometres per second. There are also many man-made objects littering space. Meteoroids and space debris are described in more detail in Chapter 9 – "Observing Satellites", which also explains how these objects are tracked. Because meteoroids and space debris travel so fast, they can cause a lot of damage to a spacecraft. Even a small hole in a critical wall could produce life threatening consequences, and so astronauts have practised mending holes in pressurized modules, in case of such a problem.

To prevent damage, the Space Shuttle is not launched when a meteor shower is expected to have particularly heavy activity. When it is in space, if it is not required to face a particular direction, it will orbit with the tail section facing toward the direction of flight. This is because the engines are no longer needed once the Shuttle is launched and so can be used to absorb the impact of any debris approaching it head on. This minimizes any damage to the heat shields at the front of the craft, which are vital during re-entry.

The ISS has bumpers that cope with the many hits that it receives by debris smaller than a few millimetres. The meteor bumper consists of a sheet of metal placed a few centimetres from the hull that acts as a shield. When a meteoroid strikes the shield, the shock of the impact is absorbed and the resulting fragments and vapour lose most of their energy, so they do not damage the station's skin. Fred Whipple of the Smithsonian Astrophysical Observatory in Cambridge, Massachusetts, proposed this type of bumper, and so it is sometimes called a Whipple shield. Particles over about ten centimetres can be detected from the ground and if any come within a certain distance of the ISS or a Shuttle, evasive action is taken. The most dangerous particles are larger than a few millimetres, but smaller than about ten centimetres. There is a 1% chance that a one centimetre size particle will pierce through the hull of the ISS in its 20 year life.

Living in Space

Freefall Effects

The terms zero gravity, weightlessness, freefall and microgravity are all used in connection with space, some of which are misleading. There is no such thing as no gravity, as might be assumed from the term zero gravity. In space, just as on the Earth, the force of gravity is always there, be it from the Earth, Moon, Sun or any other body. In a spacecraft, to all appearances there is no gravity, as everything floats. This is why it has been called zero gravity or zero G. However, everything in orbit is in a constant state of freefall, just like a stone being dropped down a well. Because the spacecraft, as well as everything in it, is falling at the same speed, everything appears to float and not drop down. Fortunately the spacecraft usually continues in this state of freefall, and does not crash into anything, unlike the stone down the well. The term microgravity is not much of an improvement, as it literally means a millionth of gravity, when what it is trying to do is explain the apparent absence of gravity, whilst hinting that gravity has not disappeared and is just being counterbalanced by the motion of the spacecraft. Microgravity is also used because spacecraft do not experience perfectly weightless conditions. In the Earth's orbit, this is due to various things, including tidal forces, caused by the change in gravity over distance. The terms weightlessness and freefall have been used throughout this book. Gravity is explained in more detail in Chapter 1 – "Introduction".

On the surface of the Earth, we are used to gravity pulling, at a rate of 1 *g*, us and anything we hold or drop, downwards, or more accurately, towards the centre of the Earth. Our hearts are used to pumping blood up to the brain, and our muscles and bones support our bodies. However, in space, the heart, muscles or bones do not need to work so hard, and the body compensates by adapting and reducing some of the superfluous material such as blood, bone and muscle. The bones do not need to be so strong and so calcium is lost from them into the blood stream. This can result in the kidneys, which filter the blood, accumulating so much calcium that kidney stones form. The muscles are not used so much, so they waste away. The volume of blood in the body is reduced because there is no longer a downward pull to distribute the fluids around the body. Instead there is a fluid shift towards the head as the fluids are redistributed to the upper part of the body and away from the lower extremities. The legs shrink and the face puffs up, which leads to blocked noses and sinuses, referred to as space sniffles. Other effects of living in weightlessness include weight loss, dehydration, constipation and space motion sickness. Because there is no downward pull to compress the spinal column, astronauts grow about five centimetres taller when in freefall, which can cause backache and nerve problems.

In order to combat these effects, extra vitamins are added to the diets of astronauts who remain in space for long periods. The amount of iron in the diet is reduced as most of the iron absorbed from food is used to make new red blood cells, but as the volume of blood is reduced, less is required and an excess of iron can cause health problems. Astronauts aboard the ISS try to minimize the effects of muscle loss and bone weakening by exercising for a few hours each day. There are three exercise machines on the ISS, these are the Resistance Exercise Device (RED) system which simulates weight lifting, a modified treadmill with Vibration Isolation System (VIS), and a modified bicycle known as a cyclergometer, with VIS. Aboard the space station *Mir*, cosmonauts wore a "Penguin suit". This was an elasticized jumpsuit with rubber bands woven into the fabric. When the cosmonauts tightened the tension on the bands, the suit simulated some gravitational effects and therefore provided exercise for the muscles. The American astronaut Shannon Lucid wore this kind of suit during her six month stay on board the *Mir* Space Station between March and September in 1996. When she returned to Earth, she was immediately able to walk, which proved the usefulness of the exercise program and suit.

Although astronauts need to be fit and healthy, NASA sees their specialist competence as a far greater importance. Therefore, NASA's astronauts are very well qualified and not always as young as might be expected. Their average age is about 41 and many do not have perfect sight. About 70% of the crew members who have participated in missions completed by NASA have needed glasses, ranging from about −5.5 to +5.5 dioptres. When in space, where weightlessness and pressure ratios affect the eyes, some astronauts perceive a change in their vision. This is mainly due to the behaviour of the liquid in the eye under freefall conditions that can change the shape of the eye. In the first few hours in space, as the bodily fluids shift towards the head, the pressure on the eye rises dramatically, which can have negative effects. However, after about 72 hours, the pressure is reduced and does not usually lead to any lasting damage.

While in space, about 90% of astronauts use optical aids. Figure 8.4 shows Mission Specialists Koichi Wakata of Japan and William S. McArthur Jr. McArthur, who is wearing glasses. During launch and landing, the astronaut experiences an acceleration force that could cause normal glasses to slip down the nose. Lightweight glasses are therefore used, weighing only about ten grams with the lenses.

FIGURE 8.4 Mission Specialists Koichi Wakata of Japan on the Left and William S. McArthur Jr. on the Right in Their Seats in Discovery for a Simulated Countdown for Mission STS–92 in 2000. McArthur is Wearing Glasses.
Image courtesy NASA Kennedy Space Centre (NASA–KSC)

When an astronaut is on an EVA, they cannot touch their face for up to ten hours. Their glasses must therefore stay on the nose and also be comfortable and not cause any pressure points. The glasses used have no hinge screws, as this reduces the risk of a loose component floating around, which could be inhaled, ingested, or penetrate and break mechanical systems. In a spacesuit, this could result in damage to the suit, and therefore a loss in pressure.

Dust and other small particles float about in space, and do not drop to the floor as they do on Earth. These particles can end up in an astronaut's eye and cause eye irritation, which can lead to inflammation. Although this is not a serious problem, the treatment is a real challenge as administering eye drops in freefall is rather difficult, as is ensuring the correct dosage.

Space Sickness

The second man in space, Gherman Titov, complained of nausea and dizziness when he was in space, although none of the early

American astronauts complained of these symptoms. However, when the larger *Apollo* craft were used, the American astronauts were not so limited in their mobility and were able to see outside. They then also began to suffer from space sickness. On the Earth, gravity affects sensors in the inner ear, which send signals to the brain. These signals are then interpreted and the body's orientation is known. Without the pull downwards, this system does not work, and everything appears very strange, and the body does not know which way is up. The nerves in the body's joints and muscles, which signal how the arms and legs have moved and what position they are in is called the proprioceptive system. The brain understands these signals under normal situations on the Earth, but when in freefall, the signals are different as gravity is not pulling the limbs down. The brain therefore has difficulty interpreting these signals and an astronaut has to look to see where their arms and legs are. The brain quickly adapts though. It learns to trust the eyes and ignore the signals that rely on gravity. The International Space Station and the Space Shuttle have been designed to help the eyes, by establishing a common sense of "up". For example, on the ISS all the lighting is on the "ceiling" of the modules and all the labels and writing face the same way. There are medications that help combat the nausea, however, these can reduce the astronaut's ability to react quickly, think clearly and recall information. About a third of all astronauts suffer from space sickness during their first few days in orbit. Some people seem to be immune to space sickness but most space visitors experience symptoms from mild headaches to nausea and vertigo. Motion sickness on the Earth does not seem to be an indicator of space sickness. Spacewalks are usually scheduled for the latter part of missions and extravehicular activity is never an option when a crewmember feels nauseous.

Toilet

Alan Shepard's sub-orbital flight in May 1961 was only 15 minutes long, and so toilet facilities were not considered necessary. However, due to delays on the launchpad he was strapped into the nosecone of the rocket for over three hours and eventually needed to relieve his bladder. As he was not able to vacate the nosecone, he asked for the electrical wires and sensors that were located all over his body to be switched off, so that he did not cause a short circuit. He then relieved himself, which was soaked up in the cotton undergarment he was wearing. As there was oxygen flowing through the spacecraft, he was actually dry by the time he launched. Facilities for such eventualities have been in place ever since.

On the Earth, when you go to the toilet, gravity assists the removal of your waste and it drops away from you. In space it does not. There are two main designs of toilet facilities in space. The first is called a body attached collector, which was used by the early astronauts and is still used on EVAs and at launch and landing. The other system utilizes the suction effect of airflow and is used on the Shuttle and the ISS toilets. The suction effect toilet was also used on *Skylab* and *Mir*.

All of the early NASA astronauts were male, and so the space toilets were only designed for men. They were adequate but not very pleasant or convenient to use. The first body attached collector was for the collection of urine only and was called the urine transfer system and consisted of a roll on rubber cuff connected to a collection bag via a valve. The cuff acted as an external catheter between the penis and the valve. Figure 8.5 shows a urine transfer

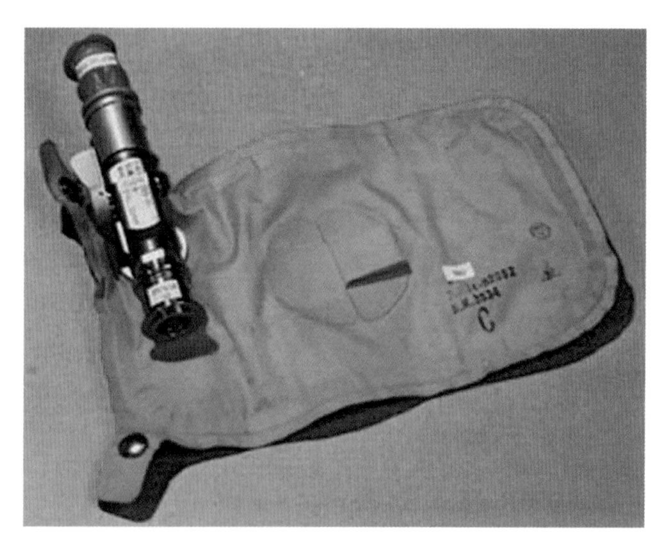

FIGURE 8.5 Apollo Urine Transfer System.
Image courtesy NASA

system. The cuff was used for one day, probably for five or six urinations, before it was replaced. Each member of the crew had their own urine transfer system, with different colour cuffs to distinguish them.

The Soviet Union had to design a female waste receptacle for their space toilets for the *Vostok 6* mission launched in June 1962, as the first female in space, Valentina Tereshkova, was onboard. She orbited the Earth 48 times and spent almost three days in space. Women did not go back into space until 1982 when cosmonaut Svetlana Savitskaya went up in a *Soyuz* to the *Salyut 7* space station. The first US female astronaut was Sally Ride in 1983. She had the luxury of the Space Shuttle toilet.

NASA's urine transfer system could be connected to an overboard dump system, which ejected the liquid into space. When a spacesuit was worn, for example, during launch, EVAs and in emergencies, a different design of urine collection device was used called the urine collection and transfer assembly, as can be seen in Figure 8.6. This again used a rubber cuff, and could be emptied through the tube either when the astronaut was still in the suit, or after he had got out.

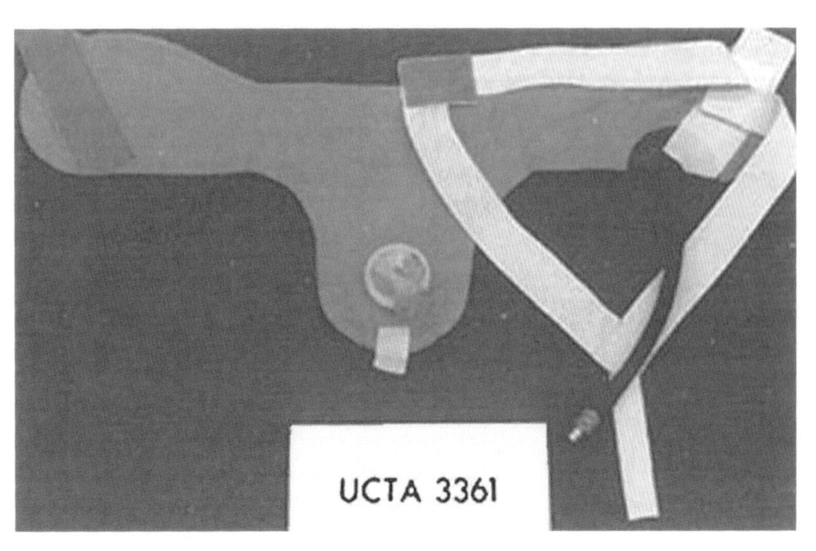

FIGURE 8.6 Apollo Urine Collection and Transfer Assembly. Image courtesy NASA

For the *Apollo 12* and subsequent missions, a device that did not require intimate contact with the astronaut was used, called the urine receptacle assembly. This was an open ended, cylindrical container that could be hand-held. This could be connected to a flexible urine dump line, which was attached to the overboard dump system. The container had a honeycomb cell insert that provided a large contact area that acted as a bundle of capillary tubes. The capillary action produced by each cell of the honeycomb tended to hold the fluid in place in the weightless environment until it could pass into the urine dump line. The cap permitted the device to be exposed to space vacuum for venting between uses. Figure 8.7 shows how the urine receptacle assembly and the urine transfer system were connected to the overboard dump system. Additional hardware enabled the urine to be stored, either for later analysis, or to prevent the glittering frozen urine from interfering with the viewing conditions through the windows. The visible frozen urine has been nicknamed the constellation Urion.

Defecation was much more complicated and unpleasant. It could take between 45 and 60 minutes to complete. The astronauts would attach a specially designed plastic bag to their buttocks with

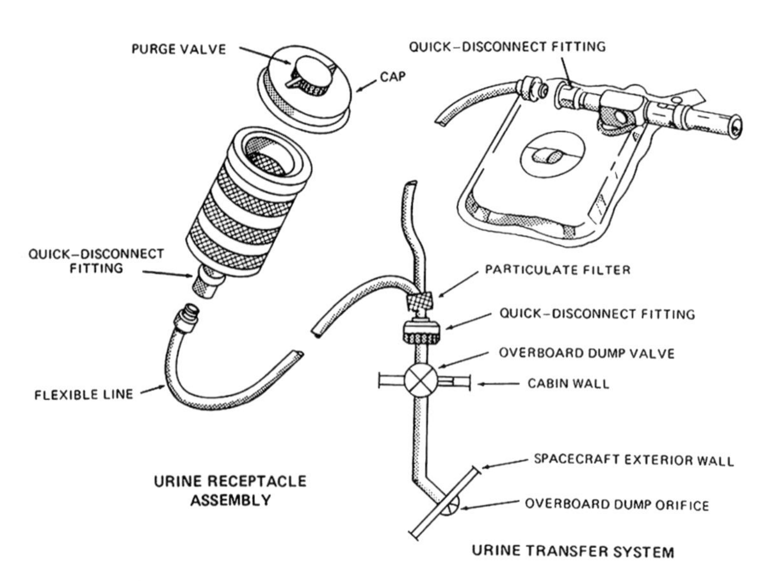

Figure 8.7 Apollo Urine Receptacle Assembly.
Image courtesy NASA

sticky tape. The bag had a finger pocket in the centre of one side, which was like a glove, so that a finger could be inserted but remain protected. After defecation, the finger was used to separate any remaining matter from the anus, and push the faeces down into the bag. After using tissue wipes, which were also put into the bag, the crewmember then had to seal the bag and knead it, so that it mixed with a germicide that prevented the formation of bacteria and gas. It was then stored in the waste stowage compartment. Low residue foods were used during flights that used this facility, and some astronauts used drugs to suppress their bowel movements.

When the astronaut was wearing their pressure suit, a faecal containment system was used in case they could not prevent defecation. This was a pair of shorts with layers of absorbent material that would contain any excreta. Improvements in technology have enabled astronauts nowadays to wear a Maximum Absorption Garment, which is like a nappy or diaper, during launch, landing and EVAs. This garment provides for hygienic collection, storage, and eventual transfer of urine and faeces. It is expected that passengers on sub-orbital space planes, such as those expected to be used for space tourism, will only have this form of toilet facility.

Space toilets onboard the Shuttle and the ISS are an improvement, although when onboard the *Soyuz* craft, astronauts try to restrict themselves to only using the urinal, by having a pre-flight enema and either fasting before the flight or eating a minimal residue diet in the preceding days.

The toilet or waste collection system on the ISS and Shuttle are very similar. They consist of a urinal and a commode. Both men and women can use the urinal either standing up or seated on the commode. The urinal consists of a flexible hose with attachable funnels. Each astronaut has their own funnel, shaped differently for men and women. Urine is drawn into the funnel and down the hose by an air flow provided by a fan. The liquid and air mixture is carried to a rotating chamber, which throws the liquid to the outer walls by centrifugal force. A blower moves the air through an odour and bacteria filter and returns it to the cabin. The liquid is then drawn into a wastewater tank. On the Shuttle, this is vented overboard, on the ISS it is put on the *Progress* vehicle for disposal. In the future, when the permanent environment and life support system is installed and activated, the waste water on the ISS will be recycled. This will purify the water after collection by removing any gas bubbles from the liquid, and then filtering it like coffee in a coffee filter. All particles with a diameter larger than 0.5 microns will be trapped in the filter. The average thickness of a human hair is about 10 microns. A high-temperature catalytic reactor assembly will then remove any remaining organic contaminants or microorganisms. After testing, if the water is not pure enough, it will be sent through the system again. The clean water, which will be cleaner than water out of most taps on Earth, will be sent to a storage tank.

On the Shuttle, the commode seat is contoured to make sure the user is in the correct position. It is made of a soft material so that it makes a good air seal with the buttocks of the astronaut. On the ISS the toilet seat is made of oak. Restraints on both the Shuttle and the ISS are used to keep the astronaut in the proper position. For urination whilst standing, the feet are slipped under a toe bar at the base of the commode. When sitting, Velcro straps are wrapped around the feet and bars are placed over the thighs. A backup method of Velcro thigh straps is also available. Handholds can be used to help position and stabilize the crewmember. Faeces are drawn downwards by air flowing in through holes under the seat. Solids and any liquid waste and toilet paper is trapped in a porous bag, but the air and other gases can flow though it and out through the filter and back to the cabin. When the astronaut has finished, they open a valve that exposes the contents of the bag to the vacuum of space. This freeze dries all of the waste and stops it smelling. After use the toilet is cleaned with wet wipes and it is also cleaned once a day with a cleanser. If the toilet system fails, astronauts have to revert back to the *Apollo* style faecal bags, which are then stored in the commode. The solid waste is later transferred to a *Progress* vessel, which is made to burn up when it re-enters the Earth's atmosphere, therefore disposing of the waste by incineration.

Ablutions

There is no shower onboard the ISS or the Shuttle, so astronauts usually take a sponge bath every day. They use two washcloths, one to wash with and the other to rinse with. They wash their hair with a rinseless shampoo, which was designed for bedridden hospital patients. Any excess water is sucked off into the wastewater tank, with a vacuum cleaner type tube. When they clean their teeth, astronauts usually swallow the toothpaste or spit it into a towel. As saliva is more concentrated in weightlessness situations, more tartar forms on the teeth, so many astronauts use chewing gum and also massage their gums to keep their mouths healthier. Although shaving cream and razors are permitted on the ISS, most astronauts prefer to use an electric razor as it is simpler. Haircuts are performed on each other with the aid of a vacuum cleaner type tube, which sucks the cut hair away.

Clothing

Astronauts wear special suits during launch and landing. If they are aboard a *Soyuz* craft, they wear the Russian Sokol suit, and if flying on the Shuttle they wear the Advanced Crew Escape Suit or ACES. Both these suits are designed to protect the astronaut if there is a pressure leak aboard the craft and can also deliver oxygen for breathing if the atmosphere is contaminated. They also have inflatable bladders in the legs, to help to prevent pooling of blood in the lower body during re-entry. If the astronaut has to bail out into a cold atmosphere or into water, the suits will also act as thermal protection. Each astronaut wears a helmet, communications cap, gloves and boots and carries equipment such as a parachute harness, parachute pack with automatic opener, pilot chute, drogue chute and main canopy, a life raft, two litres of emergency drinking water, flotation devices, a radio and beacon, signal mirror, shroud cutter, sea dye marker and flares.

Once in orbit the crew can remove their suits and wear normal clothes. Most garments are made of cotton, as this chars rather than melts at high temperatures or in flash fires. The astronauts' clothing is designed to be comfortable and functional, and must take into account the physiological changes that occur to the body when in space, such as the increase in height when the spine is no longer compressed by gravity. Aboard the ISS, shorts and trousers contain at least three pockets that can be opened and securely closed with one hand. They also have a Velcro pattern on the front of the legs to support a removable pocket. The astronauts can also wear long or short sleeved T-shirts, slippers, pyjamas and they have overalls for working in. Each crewmember also gets a pair of running shoes to use on the station's treadmill and another pair of shoes to wear when using the station's exercise bicycle. The ISS crewmembers chose their clothes months before their mission, and the clothes often arrive before the astronaut via a *Progress* resupply vessel.

To save water, there are no laundry facilities aboard the ISS. The crew do not change their clothes as often as they would on Earth but they do not get as dirty. Also, it requires very little physical effort to move about in space, and as the temperature is controlled, the body does not have to sweat to maintain a constant temperature. Some astronauts aboard the ISS do not change their underwear for three–four days at a time. Astronauts aboard *Mir* only had one clean pair a week. The astronauts do exercise daily though, but wear one pair of shorts and one shirt for every three days of exercising. They can wash themselves every day after exercising. After a garment has been worn as many times as possible, it is placed in a bag for disposal. Very little clothing is brought home from the Space Station, most of it is placed in the *Progress* resupply vehicle, along with all the other rubbish. This method of clothing supply and disposal only works because the ISS is in orbit around the Earth. If it were not able to receive regular supply visits, either enough clothes for a long duration stay would need to be aboard at the start, which would take up a lot of space, or some form of laundry facilities would be needed.

When venturing outside of a spacecraft, astronauts need a spacesuit that must provide all of the necessary environmental conditions, including removal of carbon dioxide from the astronauts' breath and protection from the Sun's radiation and from bombardment by micrometeoroids. Astronauts leaving the ISS wear either the American-made extravehicular mobility unit or EMU spacesuit or the Russian-made Orlan spacesuit. The EMU is made of interchangeable parts. The upper torso, lower torso, arms and gloves are manufactured in different sizes, and can be assembled to fit each astronaut, whether male or female, large or small. The spacesuits used in previous NASA manned space flight programs were tailor-made to each astronaut. Each part of the EMU can be reused about 25 times. The Orlan suit is a one piece suit, and is only available in one size. However, it does have various sizing straps within it, which can be used to size the suit to the individual. The arms and legs can be lengthened, and the crotch height can be changed. There are also different glove sizes available. One major advantage of the Orlan suits is that an astronaut can get into one on their own. The door at the back opens just like a refrigerator, and you can walk in. Each Orlan spacesuit can be reused, and some have been used on 15 different spacewalks. Bearings in the shoulder, arm, wrist, and waist joints allow the astronaut freedom of movement while wearing the suit, so bending, leaning, and twisting motions of the torso can all be done with relative ease. The suits are made from several layers, including a woven Kevlar, Teflon, and Dacron anti-abrasion outer layer. The materials used to construct the suit prevent fungus or bacteria growth but the suit must be cleaned and dried after each use.

As a space walk can last up to eight hours, each suit has a drink bag. A tube projects into the helmet and permits the astronaut to drink as if with a straw, whenever they like. The astronauts also wear the Maximum Absorption Garment, in case they need the toilet while they are outside. To keep the astronauts cool, the suits have a liquid cooling and ventilation system, which uses water filled tubes that flow around the body and back to the backpack, where the heat is radiated into space. The suit is equipped with a radio, so the astronaut can keep in contact with those inside the craft, and also on the ground. The astronauts wear a "Snoopy Cap" or communications carrier assembly, which includes headphones and microphones for two-way communications. They also wear a biomedical instrumentation subsystem, so flight surgeons on the ground can monitor the well-being of the astronaut.

As the astronaut is plunged into darkness every 45 minutes as they go into the shadow of the Earth, there are lights mounted on the helmet. A visor assembly is attached to the helmet, to provide protection against micrometeoroids, accidental damage and solar radiation. Adjustable eyeshades can be pulled down over the visor to provide protection against sunlight and glare.

The spacesuits are pressurized and so all joints and connections must be airtight and are made with either mechanical joints or adhesive bonding. The gloves and helmets are also clipped onto the suit, providing a pressure seal. The pressure inside the suit is less than that on the ISS, so the astronauts must breathe pure oxygen for several hours before they go out into space. This removes any nitrogen that is dissolved in the body fluids, which may be released as gas bubbles under reduced pressure, and cause the bends. If the pressure were higher, the suit would inflate and become rigid, and the astronaut would not be able to bend any joints. The cosmonaut Aleksei Leonov, the first person to walk in space, had this problem. When he tried to re-enter the airlock, he found his suit had stiffened too much and he could not enter. He had to bleed off some of the pressure in the suit, which made it easier for him to bend the joints.

To help astronauts move about while on an EVA, some form of manoeuvring unit is used. The first American manoeuvring unit was a hand held gun that fired jets of oxygen in three directions. Two jets were aimed backwards, and so propelled the astronaut forwards. The third jet was aimed forwards and acted as a brake. This unit had two major disadvantages. First, it had to be held as close to the astronaut's centre of gravity as possible, and with a bulky spacesuit, this was a matter of guesswork and trial and error. Also, ending up in a precise location was difficult to achieve and maintain, and was also physically tiring. Further developments led to the manned manoeuvring unit (MMU), which was used during the early Shuttle missions. This strapped onto the back of the extravehicular mobility unit. To move about, nitrogen gas was fed to any of the 24 nozzles arranged in clusters at the corners of the units, which gave six degrees of freedom. Using this system an astronaut could work for about six hours not attached to a spacecraft. This system is no longer used, and astronauts move around tethered to the spacecraft. However, in case an astronaut somehow becomes detached from the spacecraft, and cannot be retrieved, they carry a device called the Simplified Aid for Extravehicular Activity Rescue or SAFER. It also fits over the EMU, but when it is required, it is brought round to the front of the astronaut for easy access. It carries less nitrogen than the MMU, and so it can only be used for a limited time.

Sleep

Daylight and darkness envelops the ISS about 15 times a day as it orbits the Earth. Normal day and night time cycles are therefore disrupted, but the astronauts onboard still use the 24 hour daily cycle that we experience on the Earth. If they used a different cycle, their circadian rhythms of waking and sleeping, which are fixed in our brains and bodies, would be altered and the astronauts will feel constant jet lag. So that the sunrise does not wake them every 90 minutes, astronauts have the option of wearing a sleep mask.

Usually, astronauts sleep for about eight hours every night, and many experience dreams of weightlessness, although nightmares and snoring are not restricted to Earth-bound sleepers. As there is a tendency for weightless bodies to drift around during the night, which may result in painful collisions, most astronauts strap themselves into their sleeping bag with a waist strap. Their arms are usually free to float about during the night. As there is no up or down, astronauts can sleep in any orientation. The sleeping bag is usually attached to a wall, seat or a bunk bed inside the crew cabin. There are two small crew cabins on the ISS, each is just big enough for one person and has a large window so the astronauts can look out into space. If there are more than two astronauts aboard the station, and the commander agrees, the astronaut can sleep anywhere so long as they attach themselves to something. There are four bunk beds in the Space Shuttle, but if the astronauts are all working the same shift pattern, and therefore all sleep at the same time, those not using the bunks can sleep in the commander's seat, the pilot's seat or in a sleeping bag attached to their seats or to a wall. The bunks are over 1.8 metres long and about 0.75 metres wide, and slightly padded. The sleeping bags are attached to the bed board. One person sleeps on the top bunk, the second on the lower bunk. The third person sleeps on the underside of the lower bunk, facing the floor. There is a fourth bunk connected to the end of the triple bunks, so that the sleeper's feet are in line with the person on the bottom bunk, but their head is up by the top bunk. Figure 8.8 shows some of the *STS-68* crew in their sleep bunks. Onboard the *Soyuz*, the crew attach their sleeping bags to the curved wall of the Orbital Module.

As warm air does not rise in space, good ventilation is essential around a sleeping astronaut, otherwise they could suffocate

FIGURE 8.8 Some of the STS–68 Crew in Their Sleep Bunks.
Image courtesy NASA

in their own exhaled carbon dioxide. Ventilation fans and other equipment on board the station are quite noisy, so some astronauts wear earplugs while they sleep, but most usually get used to the noise. On the ISS, the crew are woken by an alarm clock, but occupants of the Space Shuttle and the *Soyuz* usually receive a wake up call of broadcast music from Mission Control Centre in Houston, Texas. Often, a song is chosen for a different astronaut each day, ranging from rock and roll, country and western and classical through to Russian music.

Psychological Effects

Once over any space sickness, most astronauts report an initial exhilaration at their freedom from weight. However, despite the wonderful views from the ISS, most do feel the confinement caused by a small space station, shared with other astronauts. It has been remarked that:

> All the conditions necessary for murder are met if you shut two men in a cabin measuring five metres by six and leave them together for two months.

Fortunately, the astronauts are all disciplined, highly trained people, so, so far, there has been nothing worse than a little ill temper onboard. Regular contact with home through emails and video links also helps to relieve any homesickness. For long duration flights though, such as may be used in travelling to Mars, the experiences of past crews aboard the *Mir* and ISS will need to be considered. There appear to be three distinct phases during long duration missions. The first, which usually lasts about two months, is when everyone is busy adapting to the new environment. In the second phase, there are signs of fatigue and low motivation and in the final phase, the crew become hypersensitive, nervous and irritable.

Food

In space, as everything is in freefall, liquids do not pour, food does not stay on a plate and hot gases do not rise. This leads to a few difficulties in eating. The first space travellers in the early 1960s ate their meals straight from toothpaste style tubes. These were unappetizing as the astronauts could not see or smell the food they were eating. It was also pureed, like baby food. After tubes came cubes. These were bite size, and covered in gelatine, to reduce the risk of crumbs. Crumbs in space float about and can get into instruments, as well as being unhygienic. By the time *Skylab* was in orbit in 1973, astronauts had the luxury of a dining room and table. There were footholds so the three astronauts onboard could anchor themselves and not float away. *Skylab* also had a food freezer and refrigerator, but neither the ISS nor the Shuttle have these mainly due to power and mass constraints. There are refrigeration and freezer units on the ISS, but these are used for scientific experiments and not for food. However, adding a small food refrigerator to the ISS is being considered. This will mainly be used to chill beverages.

Space food has evolved along with the space program, and the need for better tasting and more diverse foods increases with the length of time spent in space. A two week mission is literally like a camping trip and dehydrated and unappetising food can be tolerated. Astronauts onboard the ISS can eat roughly the same things as on Earth, including meatloaf, beef stew and bread pudding. It still may not look particularly appetizing, but it is the real thing.

The menu is a mixture of American- and Russian-made foods. The Russian Space Agency provides dehydrated as well as ready-to-eat meals, whereas the American food is mainly ready-to-eat, some of which is the same as that available in supermarkets. The astronauts try out each of the different foods on the Earth, and chose their own menus. However, a side effect of living in freefall is the sense of taste dulls, so even a favourite meal on the Earth can taste like a mouthful of sawdust in space. One of the most eagerly awaited foods brought up on resupply missions to the ISS are strong tasting sauces and spices, onions and garlic. Interestingly, astronauts do not seem to crave sweet, sour or salty items, just spicy foods.

The foods and menus are checked for their nutritional value, to ensure the astronauts get all of the vitamins and minerals that they need in space. These are the same ones that are needed on the ground, but some are required in different amounts, such as iron and calcium. The body usually makes vitamin D when the skin is exposed to sunlight, but as astronauts are shielded from the harmful effects of radiation, and foods rich in vitamin D, such as milk, are not suitable to be stored onboard, astronauts need vitamin D supplements.

On the ISS the meals are prepared in the Russian Zvezda service module, where there is a fold down table that can accommodate three astronauts. There are no chairs, instead, the astronauts float about as they prepare and eat their meals. There are rails on the floor, so they can slide their feet under them to stabilize themselves. The table has Velcro patches and bungee straps on it, so food containers, cutlery and any other items can be attached and not drift off. Figure 8.9 shows Expedition Seven Commander Yuri Malenchenko and Science Officer Ed Lu sharing a meal around a table in the galley onboard the ISS. Various utensils and packets can be seen attached to the edge of the table. Condiment bottles are also visible attached to the bulkhead beside Malenchenko.

The meals are packaged in single serving portions. Most Russian-made food comes in cans, which are opened with a normal can opener. The lids are not opened all of the way, so there are fewer items to float away. Once opened, the astronaut often puts a small amount of water on the bottom of the can, the very weak surface tension then holds the can onto the table, in a similar way to a rain drop that stays on a pane of glass. American food is mainly in sealed pouches, which like canned food, just needs heating and

FIGURE 8.9 Expedition Seven Commander Yuri Malenchenko on the Left and Science Officer Ed Lu Share a Meal.
Image courtesy NASA

eating. Food that needs to be rehydrated is given hot or warm water and mixed or kneaded while it is still in the packet. The packet is then opened with scissors. Most foods are eaten straight from their containers using a spoon. The moisture in the covering sauce or in the food itself is usually enough to make it stick to the spoon, as long as there are no sudden movements. After the meal, the packaging is bagged and placed in a *Progress* vessel, and therefore saves on the washing up. The cutlery and table are cleaned with wet wipes. Each food item and the cutlery is labelled with a coloured dot, so the astronauts know which items are theirs. The first Chinese taikonaut, who was only in space for 21 hours in October 2003, had a choice of Chinese-made food, including nuggets of spicy shredded pork, diced chicken and fried rice cooked with nuts and dates. These were all coated to reduce crumbs. Table 8.1 shows the menu for the first five days for astronaut William McArthur when he was Expedition 12 Commander on the ISS.

The foods are also checked to make sure they can be stored. Most of them have a five year shelf life, but they contain no more preservatives than food on supermarket shelves. The canned food and the sealed pouches are thermostabilized, in the same way as

Table 8.1 ISS Expedition 12 Commander William McArthur's Onboard Menu. Courtesy NASA

Day 1	Day 2	Day 3	Day 4	Day 5
Cottage cheese/ nuts (R)	Vegetable quiche (R)	Omelet w/chicken (T)	Mexican scrambled eggs (R)	Omelet w/chicken (T)
Chicken w/egg (T)	Granola w/blueberries (R)	Buckwheat gruel w/milk (R)	Oatmeal w/brown sugar (R)	Rossiyskiy cheese (T)
Stewed cabbage (R)	Peach yogurt (T)	Visit crackers (NF)	Applesauce (T)	Wheat bread enriched (IM)
Wheat bread enriched (IM)	Coffee w/cream (B) × 2	Apple–plum bar (IM)	Coffee w/cream (B) × 2	Peach dessert (T)
Peach–apricot juice w/pulp (R)	Orange drink (B)	Coffee w/o sugar (R)	Grapefruit drink (B)	Milk (R)
Coffee w/o sugar (R)				Coffee w/o sugar (R)
Jellied meat (T)	Split pea soup (T)	Pike perch in baltika sauce (T)	Tomato basil soup (T)	Bream in tomato sauce (T)
Kharcho mutton soup (R)	Teriyaki beef steak (I)	Pureed vegetable soup (R)	Beef steak (I)	Borsch w/meat (R)
Pork w/lecho sauce (R)	Italian vegetables (R)	Beef goulash (T)	Macaroni and Cheese (R)	Meat w/barley kasha (T)
Table bread (IM)	Candied yams (T)	Mashed potatoes w/onions (R)	Tea (B)	Visit crackers (NF)
Prunes stuffed w/nuts (IM)	Lemon–lime drink (B)	Table bread (IM)	Peach–apricot drink (B)	Apple–peach juice w/ pulp (R)
Apple–black currant juice w/pulp (R)	Tea (B)	Peach–black currant juice w/pulp (R)		Tea w/o sugar (R)
		Tea w/o sugar (R)		
Beef stew (T)	Appetizing appetizer (T)	Beef enchilada (I)	Sturgeon (T)	Shrimp cocktail (R)
Rice pilaf (R)	Pork w/potatoes (T)	Red beans and rice (T)	Lamb w/vegetables (T)	Chicken fajitas (T)
Pears (T)	Moscow rye bread (IM)	Apples w/spice (T)	Moscow rye bread (IM)	Tortillas (NF)

Candy coated chocolates (NF)	Cottage cheese/apple puree (T)	Shortbread cookies (NF)	Prunes stuffed w/nuts (IM)	Teriyaki vegetables (R)
Tea w/L & S (B)	Honey cake (IM)	Tea w/ L & S (B) × 2	Apple–black currant juice/pulp (R)	Cherry–blueberry cobbler (T)
	Apricot juice w/pulp (R)		Strawberry tea w/sugar (R)	Tea w/ L & S (B)
	Currant tea w/sugar (B)			
Peanuts (NF)	Sweet almonds (NF)	Chicken–pineapple salad (R)	Vostok cookies (NF)	Tuna salad spread (T)
Dried apricots (IM)	Grape–plum juice w/pulp (R)	Crackers (NF)	Sweet almonds (NF)	Crackers (NF)
Lemonade (B)	Tea w/o sugar (R)	Orange–mango drink (B)	Peach–apricot juice w/pulp (R)	Grape drink (B)
			Tea w/o sugar (R)	

canned food on Earth. Snack packs of dried fruits and meat are packaged in clear plastic and commercially available junk foods, such as sweets, biscuits and crackers are repackaged in single serving snack packs. Some meat products are irradiated or dosed with radiation, to improve their shelf life. A small quantity of fresh fruits, vegetables and bread products are taken into space, such as bananas, sticks of carrot and celery, bread rolls and flour tortillas. However, these must be eaten within the first few days of the flight, as they will not keep. Some items, such as fresh eggs and salads have not yet made it into space. Figure 8.10 shows some of the foods that the astronauts eat on the ISS.

Salt and pepper can be added to meals, but as they cannot be applied normally, they are in liquid form and kept in a plastic dropper bottle. The salt is dissolved in water and the pepper is suspended in oil. Other seasonings, such as mustard and mayonnaise are available in single serving sized packets.

All the drinks aboard the ISS are dehydrated and come in packets, similar to the Capri Sun™ drinks that can be bought in a

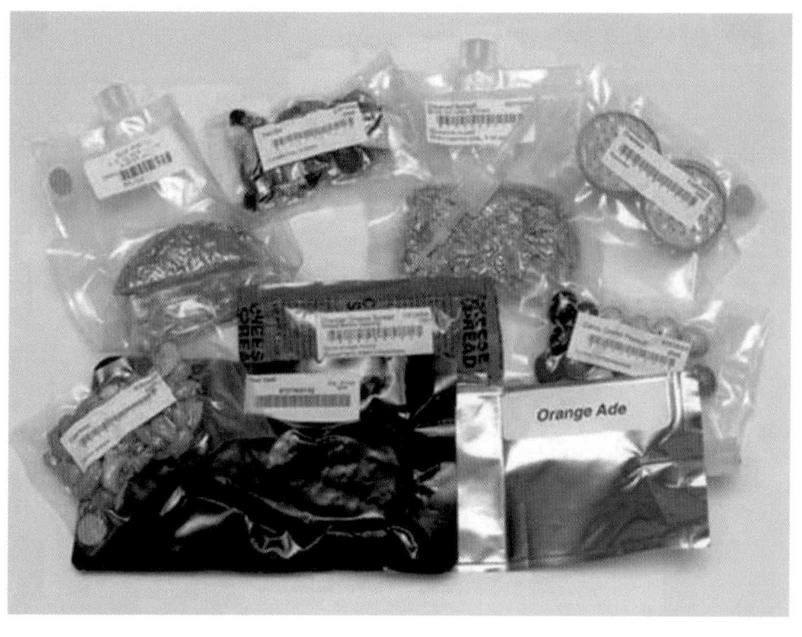

FIGURE 8.10 Assorted Bags of Snack Food and Dehydrated Food. Image courtesy NASA

supermarket. There are lots of different kinds of drinks, including juices and tea and coffee. Once water has been added, it is drunk through a straw. The straws have a clamp on them, so when the astronaut has finished drinking, but not finished the drink, the liquid stays in the pouch and does not dribble out and float around the Station. There is no cold water available on the ISS, only ambient, warm or hot water. If an astronaut wants a cold drink, they make it with warm water, and then put it in a cold area of the station, however, it never really gets very cold.

To help reduce the effects on their bodies, astronauts must be in excellent health before their mission, and return to it as soon as possible after they land. Before and after their flight, they have samples of their blood and urine analysed to indicate their nutritional status, such as bone health and the amounts of minerals and vitamins in their bodies. During the flight, they fill out questionnaires about what food they have eaten. These are emailed to the ground where nutritionists can make recommendations on ways to improve the astronauts' dietary intake.

Return to Earth

After an astronaut has returned to Earth, with the associated feeling of weight, major problems can occur. To minimize some of these effects, they drink large quantities of fluid before they land, which amongst other things, helps to restore their blood levels to those required on the Earth. The initial effect of gravity on the vestibular or balance system can cause nausea and discomfort and so returning astronauts limit the movement of their head and eyes and fix their sight on a stationary object whenever possible. In space, the heart has not had to overcome gravity, and will have become weaker and will no longer be strong enough once back on Earth. The astronaut's blood pressure will therefore drop and blood may pool in the legs. This leads to the brain being deprived of blood and oxygen and can cause faintness or even blackouts. These dangers are taken very seriously. Because of the effects of weightlessness on bones and muscles, astronauts may even have difficulty standing and most astronauts find it difficult to balance at first. However, after about three days on the Earth, the fluid levels in the body have returned to normal, but some other effects are more long-term such as the weaker bones and muscles and so returning astronauts are at a high risk of bone fractures. Once they return to Earth, astronauts exercise to strengthen their weakened bones and muscles and doctors monitor their progress. A less serious but rather disconcerting side effect of spending a long time in space has been experienced months after they have returned to Earth. Some astronauts still occasionally just let go of a mug or other object in mid air, and are surprised when it crashes to the floor.

9. Observing Satellites

When *Sputnik 1* became the first artificial satellite to circle the globe on October 4, 1957, people all over the world tuned in to hear its distinctive bleep. From certain locations, it was also possible to see the satellite as a point of light tracking across the sky.

There are now thousands of man-made objects circling the globe. These include the International Space Station, the *Hubble Space Telescope,* working satellites, dead satellites, discarded rocket stages, fragments from break-ups and other space junk. As has already been mentioned, installations around the world track about 10,000 of these objects, all of which are larger than a melon. There are probably thousands of smaller pieces that are not tracked.

Hundreds of these orbiting objects are visible to the naked eye and thousands more are visible with the aid of binoculars or a telescope. With so much orbiting the Earth, it can be quite difficult to imagine the interest that the first man-made satellite caused. *Sputnik 1* was the dawn of a new age. Radio stations broadcast the signal emitted by the satellite, while other radio stations closed down to enable listeners to tune their shortwave radio receivers to the bleep.

Although the satellite itself could be tracked from its beacon, the rocket that launched *Sputnik 1,* an Intercontinental Ballistic Missile or ICBM, was much more difficult to trace. At that time there was only one instrument in the western world that was capable of tracking it. This was Manchester University's radio telescope at Jodrell Bank, Cheshire, UK.

Construction of the 76 metre diameter telescope had begun in October 1952. Bernard Lovell had returned to the University in 1945 after working on radar during the war. He wanted to continue his pre-war research into cosmic rays by using an ex-army radar transmitter. His ideas changed and his plans grew and by August 1957, the steerable radio telescope had entered service, although still incomplete and with many teething and financial problems.

Sir Bernard Lovell says in his book *The Story of Jodrell Bank*:

> On 4th October I had no intention whatever of using the telescope to observe the Sputnik. The telescope was full of troubles, contractors were in possession, we could not drive it from the control room, there were few research cables on it, and the radar apparatus which we eventually intended to use on the American Vanguard was not on the telescope neither had we any immediate plans for installing it. We could, of course, easily have used the technology to receive the beacon transmitter from the Sputnik – the 'bleep-bleep' – but this seemed pointless because these signals could easily be picked up by a cheap commercial receiver connected to a small dipole aerial.

However, by the morning of October 7, the situation had changed dramatically. Lovell continues:

> Telephone calls from various places in London informed me that there was no defence or other radar in the country capable of detecting the carrier rocket. Although I had spent so much of my career on military radar I had been out of touch during the last few years and I simply could not believe that those responsible had so neglected the situation. Indeed, my inactivity during those few days had been founded on an assumption that many radars in the country would be detecting the carrier rocket and that any use of the telescope for a demonstration of this possibility would have been regarded with disfavour and pity by my colleagues in charge of the defence radars. Before the end of that day I had learnt with incredulity that, as least in the free world, not a single radar had succeeded in locating the carrier rocket and this was the rocket of a Russian I.C.B.M.! [sic]

He adds:

> Work which had previously been thought to take months was then completed in forty-eight hours. The change in attitude under this stimulus was fantastic.

By 6 p.m. on Wednesday, October 9, the instrument moved automatically under remote control from the control room. A few minutes after midnight on Thursday, October 10 the telescope transmitted to the Moon and received strong echoes. Lovell again:

> Echoes from the moon [sic] satisfied us that the equipment and the telescope were performing satisfactorily, but we had far to go before we could hope to achieve a radar contact from a fast moving carrier rocket whose position in the sky was only vaguely known to us.

Over the previous weekend, Lovell had agreed that there was no reason why J.G. Davies, a colleague who had worked with Lovell for ten years, should not attend a meteor conference in East Germany.

> This was just another in the series of appalling misjudgements which I made in those twenty four hours. I was to regret most deeply that I had not asked him to remain at Jodrell.

Lovell adds:

> Davies was the only one amongst us who was gifted with the ability to think instantly and intuitively about orbits. He was one of the few people in the world at that moment who, with a vast experience of meteor orbits, could have translated this experience into a judgement of the probable behaviour of the Sputnik and its carrier rocket.

Today when the positions of satellites can be found with the aid of computer software it is hard to believe that nobody could predict, with any reliability, the probable behaviour of an object orbiting the Earth at that time. Lovell:

> Until the return of J.G. Davies from Germany, J. Davis offered to help me work out the probable orbit of the rocket and direction of best look for the telescope. We did this in my office with a globe, string and map of the world.

On October 11, after many more problems, they found the rocket. Lovell recalled:

> Although there was absolutely nothing to guide us as to what an echo from a rocket would look like we were reasonably satisfied that one of the responses was at such a range and of such character that it was a response from the rocket. The next evening (Saturday 12 October) there was no doubt at all. Just before midnight there was suddenly an unforgettable sight on the cathode ray tube as a large fluctuating echo, moving in range,

> revealed to us what no man had yet seen – the radar track of the launching rocket of an Earth satellite, entering our telescope beam as it swept across England a hundred miles high over the Lake District, moving out over the North Sea at a speed of five miles per second.

By this time it seemed to Lovell that the eyes and ears of the whole world must be on Jodrell Bank especially when, on October 29 in the House of Commons, the Prime Minister (the Rt. Hon. Harold Macmillan) said:

> Hon. Members will have seen that within the last few days our great radio telescope at Jodrell Bank has successfully tracked the Sputnik's carrier rocket.

In 1987, to mark the 30th anniversary of the telescope, it was renamed the Lovell Radio Telescope in honour of Professor Sir Bernard Lovell. It remains one of the largest steerable radio telescopes in the world today.

Catalogues of Objects in Space

Of the 10,000 items in our solar system that are currently tracked, only about 700 are operational satellites, the rest is junk and other space debris. It is important to know the location of all working satellites, so that information, commands and other data can be transferred between them and the Earth. There would be no point broadcasting into space a set of commands for a specific satellite if that satellite was over the opposite side of the world. However, it also important to know where the dead satellites are, and also all the pieces of space junk, so that collisions can be avoided. Since the beginning of the space age, there have only been three collisions between tracked items. The first recognised collision occurred in July 1996, when the French *Cerise* satellite was hit by a chunk of *Ariane H-10* rocket stage. Fortunately, although a boom on the spacecraft was broken during the collision, all of its scientific payloads remained under control and the craft was able to continue its mission. The other two collisions involved a Russian non-functional navigation satellite and a piece of debris, and two other pieces of debris. The probability of a collision between two objects larger than about 10 centimetre diameter is very low. However, the International Space Station, the Space Shuttle and other satellites do take avoidance manoeuvres if there is a possibility of another object coming too near.

International Identification Number

To keep track of each man-made item in space, each piece is given a unique International Identification Number, (IIN), also called the International Designator. Before 1963 the IIN was a combination of the launch year, followed by a Greek letter to represent the number of the launch that year, followed by a number to represent which piece of that launch. When the number of launches exceeded the number of letters in the Greek alphabet, the launch number was preceded with another letter. This system soon became unwieldy and a new system was set up in 1963, and the old designators were changed. The international designation now consists of three parts consisting of the year of the launch, the number of the launch in that year and a letter or letters for each object resulting from that launch. The payload is often assigned the letter A. Any subsidiary scientific payloads in separate orbits will be labelled B, C, etc. and then inert components, for example, the burnt out rocket casing, are designated D, E, etc. The letters I and O are not used, to avoid confusion with numbers. If there are not sufficient letters to label every piece resulting from the launch, for example, if there is an explosion in orbit, then double letters are used starting with AA–AZ, followed by BA–BZ, and so on, until ZZ is reached. For example, *Vanguard 1*, which was launched in 1958 and is the oldest artificial satellite still in orbit, has the IIN 1958-002B, and the *Hubble Space Telescope (HST)* which was launched from a Space Shuttle in April 1990, has the IIN 1990-037B.

Space Catalogue

The Space Catalogue is another comprehensive list of man-made objects in space. This list includes the type of object, its orbit and where it originated. The catalogue numbers are given in sequence, as soon as the piece is detected. The Space Catalogue number and the International Identification Number do not always track together, as an item associated with a certain launch may not be detected for a long while after the first piece from that launch was detected. For example, the first module of the ISS, Zarya, was launched on November 20, 1998. This has the International Identification Number 1998-067A and the Space Catalogue number 25544, whereas a piece of debris, probably from the EVA of August 2005, has the IIN 1998-067AG and the Space Catalogue number 28792. The International Identification Numbering system considers any subsequent debris, no matter how many years later, to be part of the original launch.

As part of their space surveillance mission for the US Strategic Command (USSTRATCOM), the US Air Force Space Command, located at the Space Control Centre inside Cheyenne Mountain Air Force Station in Colorado Springs, is responsible for tracking all objects orbiting the Earth that are larger than 10 centimetres. The Space Control Centre then monitors and updates the Space Catalogue. A backup system, provided by the US Naval Space Command, is called the Alternate Space Control Centre, and would continue space surveillance if the Space Control Centre were unable to function.

As the responsibility for the Space Catalogue has changed over the years, it has been previously known as the NORAD Catalogue and the USSPACECOM Catalogue. Although USSPACECOM merged with USSTRATCOM in 2002, the catalogue is still often referred to as the USSPACECOM catalogue, or just the Space Catalogue.

Natural Bodies

The Moon is the only natural body known to be in a permanent orbit about the Earth. Meteoroids, asteroids and comets are in orbits about the Sun and fly past the Earth at high velocities. Individual meteoroids are generally too small to be tracked and designated. However, the arrival time of meteor streams, such as the Leonids in November, can be predicted, and NASA will not launch the Space Shuttle or plan a spacewalk during a meteor shower. Meteoroids are described in more detail in Chapter 1 – "Introduction".

The International Astronomical Union (IAU) is responsible for naming and updating the catalogue of natural bodies in space, such as stars, comets and asteroids – also known as minor planets. The IAU is the only internationally recognised body that has this authority and the official and permanent name of celestial bodies cannot be bought. However, the discoverer can influence the decision of the name, following various guidelines relating to where the body is located and its origin.

Minor Planets Centre and the Central Bureau for Astronomical Telegrams

As part of the IAU, the Minor Planet Centre, which operates from the Smithsonian Astrophysical Observatory, is responsible for the designation given to the minor planets in the solar system. In conjunction with the Minor Planet Centre, the Central Bureau for Astronomical Telegrams is responsible for the designation of comets and natural satellites in the solar system. The Bureau was first created in the 1880s and it is the official worldwide clearing-house for new discoveries of comets, solar system satellites, novae, supernovae and other transient astronomical events.

The Minor Planet Centre is responsible for the collection, checking and publication of observations and orbits for minor planets and comets, via the Minor Planet Circulars and associated electronic and paper publications. These are usually published on the date of each full Moon. The new numberings and names of minor planets, as well as the numberings of periodic comets, are announced in the Minor Planet Circulars.

The Central Bureau for Astronomical Telegrams is responsible for the dissemination of information on transient astronomical events, such as supernovae, via the IAU Circulars. These are a series of announcements issued as necessary in both printed and electronic forms. For example, an exploding star or supernova discovered by UK amateur astronomer Tom Boles on May 29, 2006, was first announced in the IAU Circular CBET530. When it was confirmed, it was given the official designation of 2006cr in the host galaxy UGC 8205.

Tracking Spacefaring Objects

The methods used to track objects orbiting the Earth include optical, radio and radar techniques. Optical techniques range from observations with the naked eye through to photographic methods. Radio tracking is any process that receives energy transmitted by the object being tracked. Radar bounces energy, from a source, usually on the ground, off the satellite, and back to a receiver. There are limitations to each method, and so combinations of them all are used to track satellites and space debris.

For objects in a low Earth orbit (LEO) of between about 100 and 1,000 kilometres above the Earth, the radar is a more useful method. At geosynchronous orbits (GEO) of about 35,880 kilometres above the Earth, optical telescope observations are able to detect much smaller objects than radars.

The sensors used by ground-based systems are categorised as either active or passive. Active sensors, such as radar, send out energy and read the returned signal. These usually require the satellite to carry a reflector. Satellites or spacecraft that include equipment that make them easily tracked by a specific method are called cooperative items. Such equipment usually has to be aligned towards the Earth. Dead satellites, ones whose attitudes are incorrect and space debris are called non-cooperative objects. Passive sensors are usually used to detect non-cooperative objects, although if their geometry is suitable to reflect a signal, an active sensor can still be used to track it. Passive sensors include optical telescopes that detect the sunlight reflected from the satellite.

Because of the perturbations of orbits due to such things as atmospheric drag, the gravitational forces from the Sun, Moon and the Earth and even the pressure of sunlight, the orbits of satellites and space debris must be recalculated and checked periodically. To keep track of everything man has put into space, large portions of the sky need to be observed in detail on a regular basis. Most objects are tracked on a regular basis by at least one of the many systems around the world. One of these is the Space Surveillance Network (SSN). This is a worldwide network of about twenty ground-based optical and radar sensors and one space-based sensor. The SSN has tracked more than 26,000 objects orbiting the Earth since the launch of *Sputnik 1*. Most of these have since re-entered the Earth's atmosphere. The SSN does not monitor every object continually, but does spot checks to ensure that the objects are where they are expected to be. This is due to the limitations of the network in terms of geographic distribution, capability and availability, although over 100,000 observations are still made each day. Some of the SSN sensors are dedicated, that is, their primary purpose is to perform space surveillance, others have a primary mission other than space surveillance, usually a military mission. Other systems used to track space debris from the Earth include ESA's Zeiss telescope, Germany's Research Establishment for Applied Science (FGAN) radar system and Russia's Space Surveillance System.

Optical

Optical tracking is currently used for tracking both space debris and satellites. It is also used during the launch of space vehicles. In the USA, optical tracking systems were first developed in 1956, as part of the International Geophysical Year, when the Smithsonian Astrophysical Observatory started the Satellite Tracking Program. This was a worldwide network of stations, whose aim was to obtain enough accurate photographs of satellites to be able to determine highly precise orbits. Special cameras, designed by James G. Baker and Joseph Nunn provided the tracking information. A photograph of one of these cameras, called Baker Nunn cameras, is shown in Figure 9.1. These large telescopic cameras were based on the Schmidt telescope. The Baker Nunn system imaged satellites at an altitude ranging from about 4,800 kilometres to 35,800 kilometres. The position of an object in space was determined by analyzing the star background. Twelve camera stations were established around the world, through cooperation with other countries, with one camera installed in each station. The stations were positioned in locations that had predominantly clear skies and where satellites could be continuously tracked as they circled the globe.

The Baker Nunn system suffered from two major drawbacks: First, as film was used as the detector, it had to be developed. This limited how quickly the results were available. Second, to detect satellites, personnel had to look for streaks across the photograph

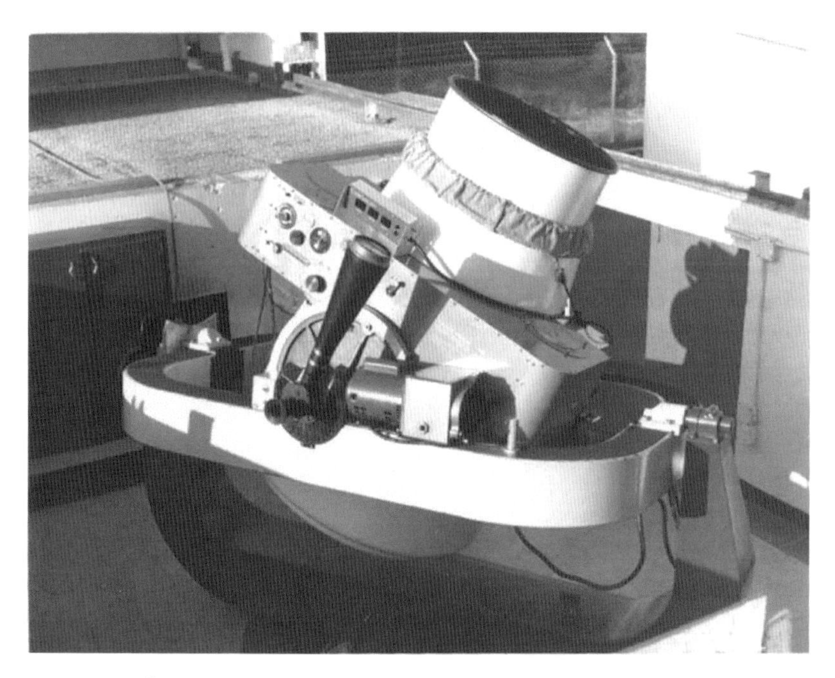

FIGURE 9.1 Baker Nunn Camera.
Image courtesy NASA

as the film had to be scanned manually. This was time-consuming and open to human error. As technology has advanced, the Baker Nunn cameras have been replaced by a new system of Ground-based Electro-Optical Deep Space Surveillance (GEODSS).

Ground-based Electro-Optical Deep Space Surveillance (GEODSS)

The GEODSS sensors are more sensitive than the Baker Nunn cameras. They can detect, image and track smaller and dimmer objects. There are three GEODSS sites controlled from the US Edwards Air Force Base, California. They are located in Socorro, New Mexico, Maui, Hawaii and Diego Garcia in British Indian

Ocean Territories. The GEODSS telescopes use low light level digital cameras and a computer instead of film. Such optical telescopes linked to digital cameras and computers are known as electro-optical sensors. The telescopes have a one metre aperture and have two special mirrors that form focused images over a 2° field of view. In comparison, the full Moon is only about half a degree wide. This type of telescope is called a Ritchey–Chretien design and is also used for the *Hubble Space Telescope's* optical system.

Each GEODSS site has three telescopes, each facing a different section of the sky. The sensors can detect objects 10,000 times dimmer than the human eye can see, down to about magnitude 16. The sensors can image objects as small as a basketball located in a geosynchronous orbit, at an altitude of about 35,880 kilometres. As the sensors are visible wavelength detectors, the telescopes can only work at night and local weather conditions and the full Moon can restrict their operation. The telescopes scan the sky at the same rate as the stars appear to move, thus keeping the stars in a fixed position in the field of view. As the telescope is scanning, cameras take rapid snapshots. Computers then overlay these images onto each other. As the stars have remained in the same position, they are easily electronically erased, leaving only items moving with respect to the star background. This will include man-made space objects, asteroids and comets, all of which appear as streaks across the image. From measurements of the streaks, the orbits of the objects in space can be calculated.

Tracking and Monitoring of Launch Vehicles

During the launch of rockets, cameras are utilized to track and monitor their progress. Since the return to flight of the Space Shuttle *Discovery* in July 2005, there are now six short range tracking cameras based on the launch pad at the John F. Kennedy Space Centre in Florida. The images from these cameras are thoroughly examined after launch, to ensure no damage has occurred to the Shuttle. There are separate dedicated cameras for the top and bottom halves of the Shuttle and also for the hydrogen vent arm above the external tank, the underside of the Shuttle's left and right wings and also the area between the external tank and the Shuttle. The short range cameras cover the launch from T-10 through to T+57 seconds, after which the Shuttle is too far away for good quality images to be produced. Seven medium range trackers are located outside of the launch pad and provide data points for the calculation of the flight path and also multiple views of the Shuttle during launch, which are also examined later for possible damage. These cameras record everything from T−7 through to T+110 seconds. Long range trackers record between T−7 and T+165 seconds and provide data points to track the Shuttle as it climbs into orbit.

Laser

Satellite Laser Ranging (SLR) is a particularly accurate method of measuring distances. It uses a laser beam which is bounced off certain satellites that carry highly efficient reflectors called retro-reflectors. The cat's eyes in the middle of a road are also a form of retro-reflector. The basic principle of SLR is very simple. A telescope tracks the satellite as it orbits overhead and fires a very short burst of laser light towards it. As each burst leaves the telescope it starts a timer. The burst travels up to the satellite where a retro-reflector bounces it back to the telescope again, and stops the timer. Since the speed of light in a vacuum is very accurately known, the time it takes for the laser light to travel through space to the satellite and back can be converted to give a distance. The speed of light changes when it passes through the atmosphere, water or any other medium. The change in the speed causes the light to bend or refract. This refraction makes a straight object appear bent when the light reflected from it passes between two different mediums, for example, when a stick is half in and half out of water. When the stick is completely out of the water, or completely submerged, it appears straight again. To take account of the varying conditions in the atmosphere, the temperature, pressure and humidity at the station are required. Models are then used to determine the speed of light through the atmosphere at those conditions and the distance to the satellite can be accurately calculated. The amount of pollution in the air has no effect on the speed of light but as dust and dirt particles are generally opaque they do block the laser light and therefore reduce the amount leaving or returning through the Earth's atmosphere.

Laser ranging is possible even when the satellite is in the Earth's shadow and during daylight hours. SLR work is carried out at, amongst other places, Herstmonceux in East Sussex, UK, where it was originally part of the Royal Greenwich Observatory's long-term research on Earth rotation and navigation. The six-man team were the only staff to remain at Herstmonceux when the Observatory moved to Cambridge in 1990. The Smithsonian Astrophysical Observatory and the USA's Department of Defence have incorporated data obtained from laser ranging into the development of their world surveying and measuring systems.

Infrared Tracking

Infrared sensors can detect objects that emit heat. As space is cold, on average about –200 °C, most objects in space are also cold. This limits the use of infrared sensors to applications such as tracking objects during re-entry.

Radio

In 1957, the USA's Naval Research Laboratory constructed a worldwide network of simple radio tracking stations called Minitrack. The system relied on a satellite transmitting radio signals that were picked up as it passed over each station. The Minitrack system was initially designed to follow the early US satellites, which used 60 milliwatt transmitters operating on 108.03 megahertz (MHz) and also 10 milliwatt transmitters operating on 108.00 megahertz. The Soviets however designed the Sputniks to broadcast between 20 and 40 megahertz, and therefore, at first, tracking the Sputnik satellites using the Minitrack system was difficult. Once the satellite stopped transmitting, the system could not track it at all. The Minitrack system has since evolved to also track non-radiating or non-cooperative satellites by reflecting signals off of them, using radar. This system is now officially known as the Air Force Fence or just the Space Fence and is described in more detail later. When the US Navy operated it, it was known as the Naval Fence or the Naval Space Surveillance (NAVSPASUR) fence. Radio signals are now mainly used for navigation and communication.

Radio Beacons

All satellites broadcast some form of radio beacon or telemetry downlink. They usually broadcast on the very high frequency or VHF band, S-band or higher microwave frequencies. Some satellites broadcast using very narrow beam antennas and therefore can only be received at specific ground stations, others, such as military satellites, only switch the broadcast on when they are above their command stations for security reasons, but many satellites transmit continuously. These broadcasts can be used to detect the direction of the satellite to an accuracy of within one arcminute. In comparison, the full Moon is approximately 30 arcminutes or half a degree wide. If the satellite is spinning, the spin rate can also be determined from the emitted signal.

Radar

Conventional radar systems use either moveable tracking antennas or fixed detection antennas. The tracking antennas steer a narrow beam of energy toward a satellite or other object and uses the returned energy to compute the location of the satellite or object. The object's motion is then followed to collect more data. A fixed detection antenna transmits a large fan of radar energy into space. When an object intersects the fan, energy is reflected back to the detection antenna, allowing the location of the object to be calculated.

Space Fence or US Air Force Fence

The Space Fence is an array of radar antennas spread across the southern USA that is used as an interferometer to make precise measurements of the path of satellites and other objects as they pass over the USA. It became operational in 1961, and works in the VHF band. Objects down to about 30 centimetre diameter can be detected with it. An upgrade to the system is planned, and it is expected that the current VHF antenna transmitter and receiver arrays will be retired and dismantled and replaced by three S-band radar systems. The S-band system will enable objects as small as five centimetre diameter, or about the size of a golf ball, to be detected in the range 160–1,000 kilometres above the surface of the Earth. The S-band system will also allow faster revisit times and, as the radars themselves will be distributed over a wider geographical area, it will give a wider view of the sky.

PAVE PAWS

The USA's PAVE PAWS is a system of ground-based ultra high frequency (UHF) phased array radars in the Massachusetts Military Reservation on upper Cape Cod, about 80 kilometres south of Boston. The UHF band covers the radio frequency range from 300 megahertz to 3 gigahertz. "PAVE" is a program name for the electronic systems, while "PAWS" stands for Phased Array Warning System. Its primary mission is to provide a warning of any incoming ballistic missiles entering North American air space, but it also has a secondary mission to track satellites.

Graves

In 2005, the French GRAVES radar system (Grande Réseau Adapté à la Veille Spatial or Radar System Adapted to Space) was commissioned. It tracks satellites in a low Earth orbit and maintains a database of orbital elements. The system transmits from the east of France, near Dijon, with the reflected signal being received in the southeast of France, about 400 kilometres away from the transmitter.

Deep Space

Tracking missions that travel out into the solar system is more complex than for near-Earth missions, as the great distances reduces the signal strength. To track and communicate with missions far out in the solar system NASA established the Deep Space Network (DSN) early in 1959. This saved each individual project from operating its own specialised network. The Russian Space Agency also has a network for distant satellites and probes, although to a much lesser degree. To maintain continuous communication throughout the Earth's rotation, the DSN has three Deep Space Communications Complexes (DSCCs) that are about 120° apart in longitude, and located on different continents around the world. There is a DSCC at Goldstone in Southern California's Mojave Desert in the USA, another near Madrid, Spain, and a third near Canberra, Australia. Each site can communicate with missions for between 8 and 14 hours, which gives a long enough overlap with the next complex to transfer the contact with the spacecraft. The three complexes are controlled and monitored by the Network Operations Control Team (NOCT) at the Jet Propulsion Laboratory (JPL) in Pasadena, California, USA. Each complex and the control centre operate all day everyday. Each complex consists of several deep space stations equipped with large parabolic reflector antennas and very sensitive receiving systems that include antennas with diameters of 70, 34, 26 and 11 metres. The data received from the spacecraft during a typical outer planet encounter can be of extremely low wattage, about 20 billion times weaker than the power required for a digital wristwatch, or about 1,000 billion times weaker than the signal received by a TV set from a commercial television station. This data is transferred via land lines, terrestrial microwave links or communications satellites to NOCT, where it is processed and delivered to Mission Control operators and mission scientists and engineers worldwide.

The DSN is mainly used as a communications link for spacecraft located far out in the solar system. This includes receiving telemetry from the spacecraft, transmitting control commands that alter the operating modes and generating the radio navigation data that is used to locate and guide the spacecraft to its destination. As well as communicating with deep space probes and satellites, the DSN can also pinpoint and track the spacecraft and identify their trajectories. The DSN was used to track the *Cassini Huygens* probe that arrived at Saturn in 2004. The time for the radio waves to reach the spacecraft at Saturn and return back to Earth is about three hours. The DSN also tracks some missions in high Earth orbits and allows a specific group of spacecraft in low Earth orbit to communicate. The DSN is also used for radio astronomy and radar observations of the solar system and Universe. It provides scientists with data about the changes in the radio signals received from a spacecraft as it passes through a planet's atmosphere or the Sun's corona. This data can then be interpreted to help scientists understand the characteristics of planetary atmospheres and the solar wind.

The Japanese space agency JAXA uses the Usuda Deep Space Centre to conduct its command operations and receive data from its deep space probes. For ESA-controlled missions, the ESTRACK network of ground stations provides the link to the operations control centre on the ground. This is a network of stations in Europe, Africa, South America and Australia owned by ESA, and additional stations in Kenya, Chile and Norway that are used by a cooperative agreement with other organizations.

Tracking from Space

The first space-based sensors designed to detect and monitor space debris and other items in Earth orbit were aboard the *Midcourse Space Experiment Satellite*, launched in 1996. For the first 18 months of its operational life, the concept of space-based space surveillance was tested, along with the accuracy and reliability of the onboard sensors. Shortly after all the tests were concluded, the infrared sensor became non operational, however the Space-Based Visible (SBV) sensor, an electro-optical camera that works in the visible light band of the spectrum, still provides useful data. The SBV sensor can detect faint objects near the sunlit limb of the Earth and also has the ability to scan large areas of the sky. Onboard signal processing reduces the amount of data produced to a manageable size. The SBV is now a contributing sensor within the Space Surveillance Network. It operates for eight hours per day and gathers as many observations as the GEODSS ground based sites, and it is considerably more accurate than the GEODSS sensors.

To replace the SBV sensor, the first of a constellation of five satellites forming a Space Based Surveillance System (SBSS) is currently being built. The constellation will operate in low Earth orbit but will look at objects up to a geosynchronous orbit, using a visible-spectrum telescope. The SBSS aims to survey an area of interest a few times a day as opposed to every few days. Another program that was intended to compliment the SBSS was the Orbital Deep Space Imager. It was to be a space telescope that would have produced high-resolution images of objects in geosynchronous orbit. However, due to budget restraints, an operational system is not currently being developed.

Amateur Tracking

If the sky is clear, a number of satellites should be visible in the early morning just before dawn or in the early evening just after dusk. It is pleasant to just watch the satellites cross the sky and daydream about what they may be, but it is easily possible to look for a specific satellite. Since the 1990s satellite predictions or tracking programs have become readily available on the Internet and provide information in various forms and are available both for viewing satellites and also for listening to their radio signals.

Optical

Although about one in ten of all satellites are above the horizon at any time, the best time to see a satellite is just before sunrise or just after sunset. At these times the Sun's rays will still be reflecting off the satellite, but they will appear against a dark background, as can be seen in Figure 9.2. The altitude and inclination of the satellite's orbit will determine when and if it can be seen from a certain place. Generally, satellites below about 800 kilometres above the Earth, including the International Space Station, will only be able to be seen once a day from a specific location.

If a satellite is in a geostationary orbit it will appear stationary and may be hard to discriminate from a star. However, all other satellites will usually appear in the sky, become brighter, move across the sky and then fade and disappear. The visual appearance and disappearance will not always occur at the

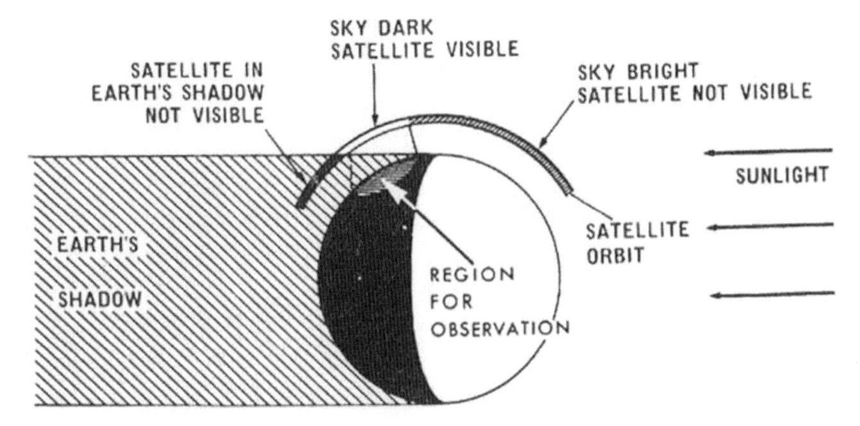

FIGURE 9.2 A Satellite Can be Seen when the Sky is Dark and the Sun Reflects from it.

horizon. When a satellite passes into the shadow of the Earth, it will become invisible, as it will no longer be illuminated by the Sun. As it comes out of the shadow, the reverse happens and the satellite will suddenly appear.

The speed a satellite crosses the sky depends on its height above the Earth. The lower a satellite is, the faster it will appear to traverse. The ISS is usually about 390 kilometres above the Earth's surface, and orbits the globe in about 90 minutes, yet it only takes between two and three minutes to pass from horizon to horizon in the sky. It is not usually visible for the whole pass, as it will enter the Earth's shadow at some stage. Meteors, or shooting stars as they are commonly called, travel much faster than satellites and can be easily distinguished.

The human eye is very effective at seeing satellites. It is capable of a field of view of nearly 180° and yet it is usually able to identify a small movement in any of the points of light that can be seen. The brightness of a satellite is referred to as its apparent magnitude, in the same way that the apparent magnitude of a star is a measure of its brightness, as viewed from the Earth. The higher the magnitude, the dimmer the star or satellite will appear. How bright a satellite appears depends on four main factors. These are the physical properties, such as the size, shape and orientation of the satellite, the surface finish, its distance away and the angle between the Sun and satellite as seen by the observer, known as the phase angle. A satellite twice the diameter of another has an effective surface area four times that of the smaller satellite. It can therefore reflect four times the amount of sunlight and appear 1.5 magnitudes brighter.

Large flat areas on a satellite, such as solar panels or antennas, can cause bright reflections that appear as sudden flashes or flares in the sky. The constellation of *Iridium* communication satellites causes the most extreme cases of these flashes. The *Iridium* Flares can be as bright as magnitude –8 for several seconds. Flares brighter than about magnitude –4 can sometimes be seen in daylight. The type of surface finish on the satellite also affects the amount of reflection. A polished metal surface reflects better than black paint or dark solar cells. The satellite's orientation affects the brightness. A rotating satellite can appear dimmer or brighter as different parts of the satellite reflect the Sun.

When a satellite is overhead it is nearer than when it is on the horizon. It will therefore appear brighter when it is seen from directly below. As described by the inverse square law given in Chapter 1 – "Introduction", sunlight reflected from a satellite twice as far away will be only a quarter as bright and it will therefore appear 1.5 magnitudes fainter. When the phase angle is small, the Sun and the satellite are almost in line, and only a small amount of sunlight will be reflected down to the observer. This is similar to observing a new Moon. Larger phase angles illuminate more of the satellite from the observer's viewpoint, and they will therefore appear brighter, just like the full Moon. Figure 9.3 shows how on the Earth we see different phases of the Moon as it orbits around us.

What to Track – Optical

Many people across the world enjoy observing satellites. There is no definitive list of which satellites to observe. Just as some astronomers want to look at the planets and solar system items and others look for galaxies, so different priorities are given to different satellites by different observers. Many observers enjoy the challenge of finding the orbital parameters of objects that are not published by the professional bodies, such as those of satellites designed to gather military intelligence, others take a particular interest in satellites

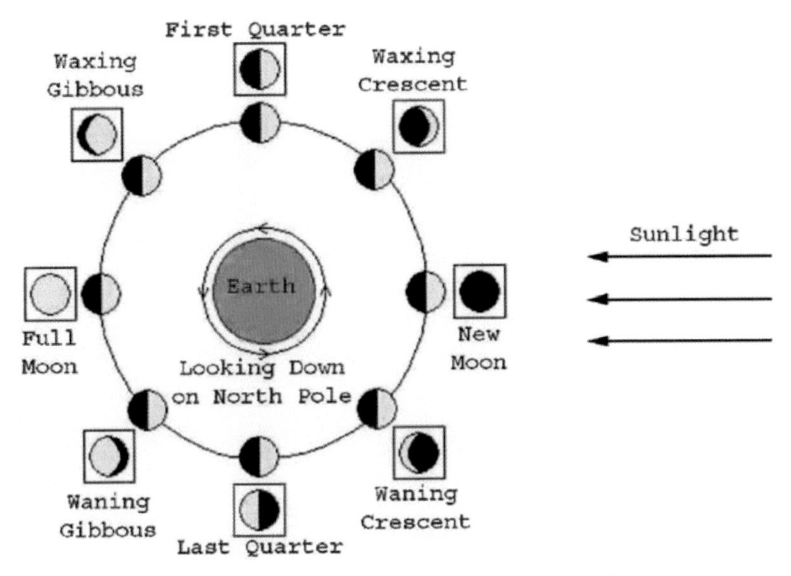

FIGURE 9.3 The Moons in the Square Boxes Show the Phases of the Moon as Seen from the Earth.

in a geostationary orbit and others aim to observe the variations in the brightness of satellites as they tumble through the sky. There are also astronomers who are seeking a new challenge. To spot a faint but rapidly moving satellite, such as the tiny *Vanguard 1*, which was America's second satellite and the oldest still in orbit, can be comparable to spotting a distant galaxy. Other challenges include transits, for example, when the International Space Station passes in front of the Sun, Moon or one of the planets.

Regular satellite observers fall into two main categories, positional observers, who precisely measure the time and position of satellites as they cross the sky, and flash observers, who measure the period of rotation of spinning satellites. The data from the flash observers are used to a provide a better understanding of the near-Earth environment, especially its magnetic field, whilst the positional observers contribute to public knowledge by finding, tracking and publishing the orbits of satellites not otherwise disclosed. Amateurs are also very good at recording the apparent brightness of both classified and unclassified satellites. This is done by comparison of the satellite with that of the stars seen in the background. Since the satellites are seen by reflected sunlight, it is possible to estimate the size of an object by noting its apparent brightness.

How to Track – Optical

A dark, clear sky is all that is needed to see many satellites. More can be seen through binoculars or a telescope. The simplest way to find a specific satellite is to get a prediction of when it is expected to appear from an Internet site. There are many sites that provide this information, including the NASA and the Heavens Above web sites. Internet addresses are given in Appendix C – "Web site Addresses". These sites provide all of the relevant information in an easy to use format for the brightest and easiest satellites to spot. The use of Internet prediction software ensures that only satellites visible from your location at a specific time are listed, so time is not wasted looking for satellites that will not appear.

The latitude and longitude of the observing site is needed to get a good prediction, as a satellite will appear at different times, from different directions or even not appear at all at some locations across the globe. For most satellites, an accuracy of about 50 kilometres or 0.5° is sufficient, however, for some items, such as the *Iridium* flares, the location must be entered to within one kilometre or 0.01°, as the path of the observed flare is very narrow at the Earth's surface. The Heavens Above introductory page provides a database of town and city names, from which geographical coordinates can be obtained. Alternatively, they can be acquired from maps, GPS or from web sites and entered manually. The information given by satellite prediction web sites can seem a little confusing at first. Appendix D – "Practical Information for Observing Satellites" explains each part of the prediction in detail. The appendix also includes detailed information on how to make positional observations, flash period measurements and also brightness measurements. Information and orbital data on satellites and space probes is available from various places but often appears in different formats. Due to name and responsibility changes, a piece of information may have several different labels from different sources. Appendix D – "Practical Information for Observing Satellites" also gives details about the different formats and labels.

Knowing the correct time is also important when trying to spot a satellite, as some satellites are only visible for a few minutes and can be easily missed. An accurate time can be obtained from an analogue radio broadcast, such as the BBC's time pips or the UK's National Physical Laboratory's time signal, known as MSF. It can also be obtained from telephone services. Care must be taken if using a digital radio broadcast or the Internet, as there is a time delay in most digital receivers and computers, therefore any time signal gained through them is not necessarily accurate. A lot of sites will give the time as Universal Time (UT), although some prediction sites will give the local time. UT used to be called Greenwich Mean Time or GMT. The UT with respect to the observing location can be found on various web sites.

The accuracy of observations depends on the person making the observations and the equipment used. The most experienced and well-equipped amateur can produce results that rival the professional and a positional accuracy of around one arcminute can be achieved. With accurate timings, observations can be good enough to calculate the orbit so that the satellite can be re-acquired the following night, having travelled nearly half a million miles. This is particularly useful for the unpublished satellites. When combined with observations made by other amateurs, the orbits calculated are accurate enough to locate the satellite after several months if it had become invisible due to daylight or eclipsed by the Earth's shadow.

Amateur Observing Community

The Visual Satellite Observers' web site provides more information about observing satellites and the formats used to share observers' data. There is also a mailing list for visual satellite observers called SeeSat-L. This allows communications among visual satellite observers and is intended to help all observers, wherever they are in the world, whatever their experience or interests. Beginners can describe their first sightings of the International Space Station and those who make positional or flash period observations, the results of which are mainly numerical, are also welcome to contribute. As the participants have diverse interests and everyone can contribute, SeeSat-L has become a valuable tool for the amateur observing community. It provides, amongst other things, reports on piloted spacecraft, rocket burns, propellant dumps, Space Shuttle water dumps, re-entries, *Iridium* flares and tethered satellites. Assistance in identifying unknown satellites is also given, along with discussions about observing techniques, prediction software and estimating orbital elements.

Other Sources of Information

The Space Track web site provides access to satellite orbital data and can be easily searched by a variety of methods, including using the satellite's common name, such as *Telstar* or *HST*, or by the launch date, International Identification Number or Space Catalogue Number. It also provides satellite decay messages, predicted decay forecasts and satellite reports. In order to access the web site, you must first register and become an approved user. Newly launched objects are listed on the Spacewarn Bulletin. This is a monthly publication and provides a list of launches and brief details of each launch. Dr. T.S. Kelso's Celestrak web site also provides a lot of satellite tracking information and you do not have to register to access it.

Radio

All active satellites transmit some form of radio signal, and with the right equipment, it should be possible to hear something from them as they pass overhead. There are some exceptions, including those that use a satellite in a geosynchronous orbit for a relay, and therefore transmit upwards rather than down towards the Earth. Listening to satellites is not restricted to early evening or dawn. If a satellite is passing overhead and broadcasting, it should be possible to receive the signal, no matter what time of day or night, or whether or not it is cloudy. However, as the human ear is not designed to receive radio signals, some form of equipment is required. Some signals can be heard using a simple "walkie talkie" or "handie talkie", often abbreviated to HT, with a rubber antenna, also known as a "rubber duck". Types of HT used for CB communication can pick up some satellites. Different satellites broadcast on different bands, for different purposes, as described in Chapter 7 – "Communication". Most amateurs listen to satellites on the UHF, VHF and HF bands although the higher C and X bands also carry a lot of downlink signals. Equipment that will monitor all of the bands all the time is very expensive. The more practical solution for an amateur is to select one or more frequency bands and a use a suitable receiver for these frequencies. Just listening to satellites does not normally require a licence, however, if you want to communicate using the satellites, one may be required.

What to Track – Radio

As with observing satellites, many different amateurs listen to different satellites or "birds" as they are often called. It is possible to receive data from the weather satellites and produce pictures of the Earth or the weather from Automatic Picture Transmission or APT signals. Some amateurs prefer to communicate with the astronauts aboard the Space Station, and others determine information about satellites that have no published data about them. From the signals received, it is possible to determine many properties of the satellite, such as when an event such as a repositioning burn took place, or what the rotational period of a satellite is. As one amateur stated:

> It's like trying to figure out something about a book, only by examining the paper and ink used and their special properties, but not knowing anything about the text in it and the language it was written in.

How to Track – Radio

With a bit of luck, it is possible to just scan the different frequencies and tune in to a satellite passing overhead. With further work and investigation, such as comparing the direction with published orbital elements, you can determine which satellite you were listening to. However, as with visual observing, it is simpler to use published tables or ephemeris, tune your receiver into the correct frequency before the published time and listen in for a specific satellite. The orbital data is given in the same format as for the observing satellites, although the frequencies are often given in separate tables. The radio amateur satellite corporation (AMSAT) provides orbital data, frequency tables and other information on their web site. The Zarya web site also provides some background to some of the missions, spacecraft and satellites and provides tracking information.

Listening to and monitoring satellite radio signals on a more serious level requires more specialised, and more expensive, equipment. Inexpensive scanners are designed to receive terrestrial signals, which are usually separated from each other by a fixed amount of about five kilohertz or more. As satellite frequencies are often only separated by 100 kilohertz, the signal from a cheaper scanner will be distorted or even not receivable. This type of scanner is usually only able to receive AM or FM signals. Some satellites do transmit at these frequencies, although many others transmit in SSB (Single Side Band) or CW (Carrier Wave) modes. However, with a bit of effort, especially with a better antenna, you should be able to receive signals from the International Space Station, the Space Shuttle, weather, navigation and some amateur radio satellite signals. More details about the equipment required and how to use it is given on the web sites of the amateur radio satellite communities, listed in Appendix C – "Web Site Addresses". Sven Grahn's web site includes more information about listening to satellites and also provides recordings of various satellite signals.

Amateur Radio Satellite Communities

There is a web site and mailing list for amateurs who listen to and monitor satellite radio signals, called HearSat. The mailing list allows information about receiving radio signals from low Earth orbiting satellites to be shared. HearSat encourages any amateur to post reception reports, questions and discussions about satellite signal reception and tracking. The list is also used to distribute orbital information and satellite status reports. Amateurs who focus on listening to and contacting the International Space Station have their own web site, the ISS Fan Club, which is listed in Appendix C – "Web Site Addresses". The site is not affiliated to any official body or organization and provides information about the International Space Station. Most of the content is related to amateur radio activities, as most members are active amateur radio operators. The Organization for Weather Satellite and Earth Observation Enthusiasts has a web site that encourages amateurs to receive data from weather and Earth imaging satellites and it helps amateurs share their skills with others. There is also a group who listen for meteors and are part of the International Meteor Organization. Many of these communities may have members in your location, who are willing to share their experiences and demonstrate their equipment.

Radar

Radar transmission is not particularly feasible for an amateur. However, observers located near the Space Fence in the southern USA or the French GRAVES system, should be able to pick up a radar return from anything that crosses the fence, if they have a receiver tuned in to the correct frequency.

10. Where to Go

Going anywhere in space involves crossing enormous distances. The Moon is about 385,000 kilometres away and took the *Apollo* spacecraft over three days to get there. The *Mars Reconnaissance Orbiter* took about seven months to reach Mars. As everything is moving in relation to everything else there are times when an object is closer and therefore easier to reach than other times. Missions to objects in our solar system must therefore be well planned. Most missions outside of a low Earth orbit are for scientific research, and aim to provide answers to many of the questions we have about solar system objects and the history and formation of the solar system.

The distance between the Sun and the Earth is about 150,000,000 kilometres. This distance is also known as an Astronomical Unit or one AU. The heliopause is thought to be about 100 Astronomical Units away from the Sun, and no man-made craft have yet crossed this boundary. Proxima Centauri, which is the next nearest star to us, is over 266,000 Astronomical Units away. Light from our Sun takes about eight minutes to reach the Earth and 4.2 years to reach Proxima Centauri. With current technology, spacecraft can only travel at a small fraction of the speed of light, and so it would take thousands of years to reach another star and currently there are no missions planned to explore other star systems.

The Sun

The Sun is a lot closer to the Earth than any other star, and so we know more about it than any other star. However, it is only an average star. It is not particularly large, old or hot, but neither is it particularly small, young or cool. The Sun is about three quarters hydrogen, with helium making up almost all of the remainder, but there is also oxygen, carbon and iron present and smaller traces of neon, nitrogen, silicon, magnesium and sulphur. The Sun is 108 times wider than the Earth and over 330,000 times the mass. The gravity on the Sun is almost 28 times more than on the surface of the Earth. As seen from the Earth, the Sun is magnitude –26.7.

The Sun gives off heat and light because millions of tonnes of hydrogen within it are burnt to release energy every second. The Sun does not burn in the same way that a log burns on a fire but it burns by nuclear fusion, which is a process that does not require oxygen. The intense gravitational forces and heat inside the core of the Sun changes the hydrogen into helium in a nuclear reaction that releases enormous amounts of energy. As so much hydrogen is burnt every second, the longer the Sun shines the lighter, or less massive, it becomes. The Sun has probably converted about half of its hydrogen to helium over the last five billion years, and should have enough hydrogen to burn for at least another five billion years.

There is a flow of material coming out of the Sun called the solar wind. This is composed of charged particles that flow out past all the planets and form the heliosphere. When the solar wind blows steadily the effects are hardly noticed on the Earth but it has a large effect on the tails of comets and even has a measurable effect on spacecraft trajectories. At times, large amounts of matter are ejected from the Sun. These can produce harmless effects like the aurora or northern lights, or more disruptive events such as power blackouts. These emissions are usually caused by solar flares and by Coronal Mass Ejections (CMEs), both of which happen where there is magnetic activity in the Sun. However, magnetic activity and emissions are not directly related, in the same way that a volcano and an earthquake are both areas of geological activity on the Earth, but are not directly related to each other. At the Sun's equator it takes about 25 days to make one revolution, but in the polar regions it takes over 30 days. Therefore the Sun's magnetic field lines get twisted and so the atmosphere looks like a plate of spaghetti with hot gases trapped in magnetic webs. Occasionally the magnetic fields get so tied up in knots they break like an elastic band and pump the excess energy into the gases. This brings the active regions to very high temperatures and particles are accelerated out into space. Solar flares and CMEs can be a radiation hazard to astronauts and also to high-flying planes such as Concorde, when it flew. They can disrupt satellite operations and communications systems and also cause radio blackouts by causing disturbances high in the Earth's atmosphere.

The surface of a star is called the photosphere, and the temperature of the Sun's photosphere ranges from about 4,000°C to 6,000°C. Curiously, the atmosphere above the photosphere, called the chromosphere is about 26,000°C hotter than the photosphere, and the outermost layer of the atmosphere, called the corona, is at about one million °C.

Spacecraft have never been closer than two thirds of the way to the Sun, which is a little closer than the orbit of Mercury. In March 1975, *Helios 1*, the joint mission of the Federal Republic of Germany and NASA, was the first spacecraft to get this close, and was used to study the Sun's properties, such as the solar wind, magnetic and electric fields, cosmic rays and also dust particles in space, called cosmic dust. To help protect the spacecraft, it was covered in optical mirrors that reflected about 90% of the Sun's heat. The spacecraft also spun once every second to help distribute the heat evenly. The amount of solar activity, such as solar flares, fluctuates in an 11-year cycle, called a solar cycle. *Helios 1* was launched when the cycle was at its minimum and therefore the threat of a solar flare damaging the spacecraft during its first few years of operation was quite low.

There are a few spacecraft currently studying the Sun, including NASA's *Ulysses* spacecraft, which is in orbit around the Sun, but passes over its poles rather than round the equator. ESA and NASA's *Solar and Heliospheric Observatory (SOHO)* uses a battery of instruments to study the Sun from the Lagrangian Point 1, which is described in Chapter 4 – "Movement in Three Dimensions". This is only about a hundredth of the distance to the Sun. *SOHO* is a collaboration of over 30 institutions from around the world. Other missions, such as NASA's *STEREO* and Japan's *Hinode (Solar B)* complement *SOHO*, and provide more information for teams around the world to try to understand the mechanics of the Sun. Future missions, such as the *Solar Dynamics Observatory*, will be used to give us a better understanding of the Sun's influence on Earth and the area of space near the Earth.

Near the end of the Sun's life, in about five billion years, the hydrogen within the Sun will be nearly all used up. Gravity will then cause the helium-rich core to contract and heat up. The remaining hydrogen will form a shell around the outside of the helium core and this shell will then burn by nuclear fusion and cause the gases in the shell to expand. The Sun will therefore swell up and engulf the Earth. The Sun will then be a type of star called a red giant, similar to the star Aldebaran in the constellation Taurus.

Eventually the Sun will shed its hydrogen shell and the core will shrink and become what is called a white dwarf star. White dwarfs are about the size of the Earth, but are extremely dense. One teaspoon of white dwarf could weigh over 5,000 kilograms. A white dwarf does not generate energy and it just slowly cools. As white dwarfs do not shine brightly, they are difficult to see in the night sky. The brightest star in the sky, Sirius in the constellation Canis Major, has a white dwarf companion star that orbits it every 50 years. This star is nicknamed the Pup, as Sirius is also known as the Dog Star. The outer layers of the Sun will eventually turn into a planetary nebula, which has nothing to do with planets. The nebula will last another 100,000 years or so, during which time most of the star's mass will be returned to interstellar space. The Ring Nebula (M57), in the constellation Lyra is a planetary nebula from a dead star.

Planets and Their Satellites

Planets in our solar system orbit the Sun and do not emit light. The planets that are composed of rock and metal are known as terrestrial planets. These are Mercury, Venus, Earth and Mars. Those mainly composed of liquid and gas are called the gas giants, these are Jupiter, Saturn, Uranus and Neptune. Table 10.1 shows the different properties of the planets in the solar system.

Mercury

Mercury is the closest planet to the Sun, with an orbit of about 0.38 Astronomical Units. However, the orbit of Mercury is very elliptical, or eccentric, and the distance between it and the Sun can vary from 0.31 Astronomical Units to 0.47 Astronomical Units. The Earth orbits the Sun at about 30 kilometres per second, and as the orbit of the Earth around the Sun is almost circular, the difference in speed between the closest point on the orbit to the Sun (perihelion) and the furthest point

Table 10.1 Planet and Dwarf Planet Fact Sheet

Body	Distance from Sun (AU)	Distance from Sun (km) 10^6	Length of day (Earth hours)	Orbital period (Earth days)	Diameter (km)	Mass (kg)	Axial tilt (Degrees)	Orbital inclination (Degrees)	Gravity (m/s/s)	Escape velocity (m/s)
Sun	–	–	609.1[a]	–	1,392,000	1.99E + 30	–	–	3.70	617.50
Mercury	0.39	57.9	4,222.6	88	4,880	3.30E + 23	0.01	7.00	8.90	4.30
Venus	0.72	108.2	2,802.0	225	12,100	4.87E + 24	177.40	3.40	9.80	10.40
Earth	1	149.6	24.0	365	12,750	5.97E + 24	23.50	0.00	1.60	11.20
Mars	1.52	227.9	24.7	27	6,800	6.42E + 23	6.70	1.90	3.70	5.00
Jupiter	5.2	778.3	9.9	687	143,000	1.90E + 27	25.20	1.30	23.10	59.50
Saturn	9.55	1,429.4	10.7	4,331	120,500	5.68E + 26	3.10	2.50	9.00	35.50
Uranus	19.22	2,875.0	17.2	10,747	45,100	8.66E + 25	26.70	0.80	8.70	21.30
Neptune	30.11	4,495.1	16.1	30,589	49,500	1.02E + 26	97.80	1.80	11.00	23.50
Pluto	39.54	5,870.0	153.3	59,800	2,400	1.25E + 22	28.30	17.20	0.60	1.10
Ceres	5.534	825.0	9.1	90,588	960 × 932	1.02E + 03	122.50	10.58	0.27	0.64
Eris[b]	67	10,022.3	>8 h?	203,500	1,200	1.66E + 22	?	44.00	?	?
Moon	–	0.385[c]	708.7	27	3,475	7.30E + 22	6.70	5.10	0.17	2.40

[a] Sidereal rotational period at 16° latitude

[b] Details about Eris are only known approximately

[c] Distance from Earth

(aphelion) is about one kilometre per second. As Mercury's orbit is very elliptical, the difference between the speed at its perihelion and aphelion is just over 20 kilometres per second, with an average speed of about 48 kilometres per second. Mercury's orbit has an inclination of 7°, which is by far the largest inclination of the planets. This means that it orbits around the Sun at an angle to the Earth's orbit, called the plane of the ecliptic. Mercury was one of the planets known to the ancient world, along with Venus, Mars, Saturn and Jupiter. The *Epic of Gilgamesh* is thought to refer to the planet Mercury. It is one of the oldest written stories ever found and written in cuneiform script on 12 clay tablets. The tablets date from about 2150 BC. As Mercury is so close to the Sun, from the Earth it appears as an evening star just after sunset or a morning star just before sunrise, but is not seen in the middle of the night. It is so close to the Sun that it cannot be imaged by the *Hubble Space Telescope*. Mercury is the smallest planet in the solar system, now that Pluto is no longer classified as a planet. If the Earth were the size of a grapefruit, Mercury would be about the size of a golf ball and the Earth's Moon about the size of a Brussels sprout. Mercury itself does not have any moons.

When Venus is on the other side of the Sun to us, Mercury is our closest neighbour. However, we still know relatively little about the planet. Most of what we do know has come either from Earth-based observations or from the probe *Mariner 10*, which flew by Mercury three times in 1974 and 1975, but only managed to map just less than half of the surface. Up until 1965, it was thought that Mercury kept the same face towards the Sun in the same way that the same face of the Moon always points towards the Earth. However, radar observations have shown that Mercury rotates three times in every two of its years. Its year lasts 88 Earth days. Although Mercury rotates just over once every 58 days, because of its motion around the Sun, its day is 176 Earth days long, in the same way that the Earth takes a little over 23 hours and 56 minutes to rotate on its axis, but its day is 24 hours long.

Viewed from the surface of Mercury, the Sun would look between about two and three times as large as it does from the Earth, depending where it was on its orbit around the Sun, and it would be up to 11 times brighter. Mercury has an extremely thin atmosphere but for most purposes this is deemed negligible. There is no evidence to suggest that the atmosphere was ever much thicker and so it is thought that there has never been any surface erosion from wind or water. The temperature on Mercury ranges from about −183 °C on the night side of the planet to about 427 °C on the sunlit side. If the planet had a substantial atmosphere, convection currents and wind would even out these temperatures.

The surface of Mercury looks very similar to the surface of the Moon. There are many impact craters where large objects such as meteoroids and asteroids have struck the surface. One of the largest craters is the Caloris Basin, which is about 1,300 kilometres in diameter. Mercury also has regions of relatively smooth plains. It is much more dense than the Moon, and the second densest planet in the solar system. Only the Earth is denser. Mercury also has a magnetic field that is about 1% the strength of the Earth's. The density and the magnetic field imply that Mercury has a large iron core and a relatively thin mantle and crust.

As the planet is so close to the Sun, and it has an inclined and an eccentric orbit, missions to Mercury are technically difficult. The extremes of temperature are demanding to any spacecraft design and any solar flares could prove lethal both to electronics and anything living, unless they were well protected. The useful operational life of spacecraft operating in these conditions is often much less than those operating in less harsh environments, and to reduce the probability of damage by a solar flare, spacecraft that approach the Sun are often launched near the solar cycle minimum.

To pass near to Mercury, a spacecraft leaving the Earth has to be slowed down considerably so that its perihelion distance from the Sun is reduced. A rocket that could carry the amount of fuel required to do this by only using a propulsion system would either be prohibitively expensive or even impossible using current technology. The Italian mathematician and engineer Giuseppe (Bepi) Colombo suggested a method to determine when to launch the spacecraft *Mariner 10* so that it could use a Venus fly-by to slow it down and to get it into a solar orbit that would bring it back repeatedly to Mercury. This method proved successful, although *Mariner 10* only flew by and did not enter into an orbit around the planet.

In the 1980s a Jet Propulsion Laboratory scientist, Dr. Chen-Wan L. Yen, suggested a trajectory that used multiple fly-bys to reduce the spacecraft velocity to reduce the orbit around the Sun enough to allow insertion into a Mercury orbit, with a sizeable payload. NASA's *MESSENGER* spacecraft (Mercury Surface, Space Environment, Geochemistry and Ranging) uses such a trajectory. *MESSENGER* was launched in 2004 and after fly-bys of Earth, Venus, and Mercury to bring it closer to the Sun, it will enter Mercury orbit in March 2011. The mission will, among other things, study the chemical composition of Mercury's surface, the history of its geology, the nature of the magnetic field and the magnetosphere and the size and composition of the core. It will also try to determine whether there is water ice in the shadows of some craters at the North Pole. Earth-based radar observations have suggested that this is possible, and as Mercury is not tilted on its axis like the Earth, it does not get seasons in the same way and the poles will never get much direct sunlight. Most information will be gathered from the northern hemisphere, as the orbit around the planet will be highly elliptical. The spacecraft will probably pass less than 1,000 kilometres above the North Pole but will be about 15,000 kilometres over the South Pole. By studying the planet, we may also learn more about planetary formation and magnetic field generation.

Another planned Mercury mission is the joint ESA and Japanese *BepiColombo* mission. This is due to launch in 2013 and will take six years to get into orbit around Mercury. It will then spend about one Earth year in orbit investigating the planet. After it is launched into a geostationary transfer orbit around the Earth, the orbit will be increased using chemical propulsion, until it can take advantage of a lunar fly-by and start on an interplanetary trajectory. To reduce the amount of propellant required during the interplanetary trajectory, an ion propulsion engine and five gravity assists, one from the Earth, two from Venus and two from Mercury will be used. On arrival at Mercury, the ion propulsion engine will be jettisoned and the spacecraft will enter Mercury's orbit by using chemical propulsion, which can slow it rapidly unlike the ion propulsion engine, which can only provide small but continuous propulsion. The spacecraft and the scientific equipment onboard will be protected with high temperature resistant thermal protection and it will use solar arrays to produce electricity.

There are no plans for missions with landers or manned craft to visit Mercury in the near future.

Venus

Venus is the second closest planet to the Sun and it is also the planet that comes closest to the Earth. Venus is about 0.7 Astronomical Units away from the Sun and when they are closest, the distance between the Earth and Venus is about 0.27 Astronomical Units. For comparison, the distance between the Earth and the Moon is about 0.0026 Astronomical Units. As Venus' orbit is very nearly circular, its distance from the Sun does not change significantly over a Venusian year, which is 225 Earth days long. As the planet spins in a retrograde direction the Sun rises in the west and sets in the east, and its day is just over 116 Earth days long.

From the Earth, Venus is the third brightest object in the sky, after only the Sun and the Moon. Venus has an atmosphere and very thick clouds hide the surface of the planet from view. These clouds, and the closeness to the Sun, give the planet its high albedo, which is why it shines so brightly, with a maximum brightness of magnitude –4.6.

As Venus is an inferior planet, that is, it orbits closer to the Sun than the Earth does, it shows phases when viewed from the Earth, just like the Moon. Mercury also shows phases, but these are difficult to see with an Earth-based telescope. As it is closer to the Sun than we are, Venus appears as an evening star or as a morning star. It can even be seen during daylight hours, and, when it is at its brightest, it can cast a shadow on clear, dark nights that are not subjected to light pollution. Venus is often called Earth's sister planet, as it is very similar in size, at about 95% of the Earth's diameter, density and chemical composition and the gravity on the surface of Venus is just a little less to that on Earth. However, Venus does not have any moons.

Because of these similarities, it was thought that Venus may harbour life, but further scientific investigations have revealed that there are some very significant differences between the two planets, which may make Venus the least hospitable place for terrestrial life in the solar system. The atmosphere is mainly carbon dioxide, but it is so dense that on the surface there is a pressure 90 times that of the surface of the Earth. To find the same pressure on the Earth, you would need to dive down to almost one kilometre underwater. The clouds are composed of droplets of sulphuric acid and other chemicals and, over about 60 kilometres above the surface, they are about 30 kilometres thick.

The dense carbon dioxide atmosphere has caused Venus to heat up by a runaway greenhouse effect, so the average temperature on the planet is over 450 °C. Lead is a liquid at this temperature. Because of this greenhouse effect, Venus is hotter than Mercury, although it is nearly twice as far away from the Sun. The atmosphere and the winds ensure that there are no extremes of temperature on Venus and the planet stays at about the same temperature during the night as during the day. The wind speed on the surface of the planet would be classified as a gentle breeze on the Earth. However, above the cloud tops the wind races at about 400 kilometres per hour. The jet streams above the Earth usually vary between about 100 and 200 kilometres per hour.

As mentioned previously, the first fly-by of Venus was by the Soviet Union's *Venera 1* spacecraft on February 12, 1961. The USA's *Mariner 2* did a Venus fly-by in December 1962, and showed the surface to be about 425 °C. In 1967 *Venera 4* was the first probe to be placed directly into the atmosphere and to return atmospheric data, although the pressure on Venus crushed it before it reached the surface. It was not until 1975 that a probe successfully landed on the surface of Venus. It was the Soviet Union's *Venera 9* and it survived for 53 minutes and transmitted the first black and white images of the planet's surface.

Other orbiter, fly-by and lander missions have sent back data about Venus. The USA's *Magellan Orbiter* arrived into Venus orbit in 1990, having been released into Earth orbit from the Space Shuttle. During the next four years it mapped 99% of the planet's surface using a radar imaging system. ESA's *Venus Express* Obiter entered Venus orbit in April 2006 and is currently investigating the Venusian atmosphere.

In 1967, Edward A. Willis, Jr. of the Lewis Research Centre, Cleveland, Ohio produced a technical memorandum for NASA for a manned Venus orbiting mission. This studied the feasibility of a manned orbiting stopover round trip to Venus, with departure dates between 1975 and 1986. He concluded that a Venus mission could be accomplished using *Apollo*-level technology, the same technology that got Man to the Moon, by using a highly elliptical orbit around the planet. The effect of radiation exposure on the astronauts was not considered in the report. Only an orbiting mission was considered at the time and not a manned landing because so little was known about the surface conditions of Venus, and also the nearly Earth-strength gravity made a manned landing and take-off system extremely difficult and heavy. His report also states:

> It is interesting to note that contemporary scientific opinion does not reject the possibility that some form of life could have developed on Venus, and that there may be suitable sites for a manned landing.

As spacecraft probes sent back more information about the planet, a manned Venus trip was deemed unfeasible, and unmanned spacecraft have always been used. There is no evidence of life being able to survive on Venus and no current plans for a manned mission.

Earth

The Earth is the third planet from the Sun, at a distance of 1 Astronomical Unit. The atmosphere is about three quarters nitrogen, almost one quarter oxygen with small amounts of carbon monoxide, carbon dioxide and water vapour. The average temperature on the surface of the planet is about 15 °C. As water covers almost three quarters of the surface and there are clouds in the atmosphere, from space the planet appears blue and white. If the water were removed it would look quite similar to the other rocky planets and satellites in the solar system. The core of the Earth is solid in the middle with a liquid outer layer and is composed of iron and nickel. Above the core is the mantle, comprising of silicate materials. The crust or skin of the planet can be divided into two different types, the continental crust and the oceanic crust. The continental crust is up to 70 kilometres thick and is made of silicate rocks that are rich in aluminium, silica and calcium. The oceanic crust has an average thickness of about Eight kilometres and is composed of silicate materials that are rich in iron and magnesium.

The first spacecraft to circle the Earth was *Sputnik 1* in October 1957. By March 2007 there were over 3,000 payloads orbiting the Earth and over 8,000 spent rocket parts and debris. The first manned Earth orbiting space station was the Soviet Union's

Salyut 1, which was launched in April 1971. This and following space stations have enabled scientists to understand some of the effects of long duration space missions on the human body and also to undertake experiments under weightless conditions. In a low Earth orbit, astronauts aboard a space station are protected by the Earth's magnetic field from much of the harmful radiation from the Sun and so all space stations so far have travelled in a low Earth orbit.

The Moon

The Moon is about 0.0026 Astronomical Unit or 385,000 kilometres from the Earth and orbits it in 27 days 7 hours and 43 minutes. It takes the same time for the Moon to orbit the Earth as it does to rotate once on its axis, as the Moon is tidally locked to the Earth, as described in Chapter 2 – "Rockets and Spacecraft". This means that the same side of the Moon always faces the Earth and we never see the far side. Due to libration, or a slight wobble of the Moon as seen from the Earth, we actually get to see 59% of the Moon's surface at different times. Because the Earth also rotates and is also moving around the Sun, it appears to us on the Earth that the Moon takes 29.5 days to orbit the Earth.

Because of the effect of gravitational forces between the Earth and the Moon on the oceans and crust, the Earth slows down by about 1.5 millisecond per century, making the day slightly longer, and the Moon rises to a higher orbit by about three centimetres every year. It is not known exactly when or how the Moon formed, although it is thought to have been created about 4.6 billion years ago from the ejected material from a collision between the Earth and an object about the size of the planet Mars.

The Moon is just over a quarter of the Earth's diameter and it is 1/80th of its mass and is by far the brightest object in the night sky. The Moon is the fifth largest satellite in the Solar System, and as it is comparatively large compared to the Earth, some planetologists view the Earth and Moon as a binary or double planet system.

From the Earth with the naked eye, two different types of area can be seen on the Moon's surface. These are the bright highlands or terrae and the darker areas, which are called maria or seas, although they are completely waterless. The maria are huge impact craters that were later flooded by molten lava. The surface of the Moon has many impact craters from asteroids and meteoroids and it is mostly covered in regolith, which is a mixture of fine dust and rocky debris produced by the impacts. The Moon probably has a small core, above which are the mantle and then the crust. The core is thought to be iron rich and about 800 kilometres in diameter. The mantle makes up about 90% of the volume of the Moon. The crust varies in thickness from almost non-existent to over 100 kilometres. Oxides are the most abundant chemical on the Moon but there are also a lot of silicates. Other chemicals, such as magnesium, iron and titanium are also present, as are aluminium and calcium and traces of sulphur, phosphorus, carbon, hydrogen, nitrogen, helium and neon. As the Moon has virtually no atmosphere and no magnetic field, the surface is exposed directly to the solar wind. From this solar wind hydrogen and helium are trapped in the surface layers of the Moon and one isotope of helium, helium 3 is found in abundance there, although it is rare on the Earth. It is thought that helium 3 could be used for nuclear fusion energy, although this technology has not yet been developed fully.

The first spacecraft to visit the Moon was the Soviet *Luna 2* in 1959 and the first photographs of the far side of the Moon were obtained from the Soviet spacecraft *Luna 3* later that same year. Between July 1969 and December 1972, after a journey of nearly four days, six *Apollo* missions landed astronauts on the Moon. As the Moon's gravity is about one sixth of that on the Earth, it is easier to leave the surface of the Moon than it is the surface of the Earth. This enabled small launch vehicles to be used to return the astronauts from the lunar surface.

As the Moon has not had the weathering effects of an atmosphere, water or life, clues to its formation are more easily seen than on the Earth. The more that is known about the Earth and the Moon, the more we can understand other planets and therefore planetary science improves. Each *Apollo* mission brought back rock samples that were collected from different regions. Figure 10.1 shows the legs and feet of *Apollo 12* Commander, Astronaut Charles Conrad Jr., as he uses tongs to collect lunar rock samples. The Soviets also returned samples of rock with their automated spacecraft *Luna 16*, *Luna 20* and *Luna 24*. The data obtained from these rocks has been correlated with other data produced from

FIGURE 10.1 Close-up View of a Set of Tongs Being Used by Astronaut Charles Conrad Jr., to Pick Up Lunar Samples During the Apollo 12 Extravehicular Activity.
Image courtesy NASA

missions such as *Pioneer*, *Ranger* and *Zond*, all of which has added to our knowledge of lunar science. Other information, including data from temperature sensors, spectrometers, laser rangers and magnetometers has been obtained from NASA's *Clementine* mission in 1994 and *Lunar Prospector* in 1999 and ESA's *SMART-1* mission, which entered lunar orbit in 2004.

When the telescope was invented in the early 17th century, the Moon was an obvious target for investigation and, up until the middle of the 20th century, visual observations and hand drawings were used to map the visible surface of the Earth's only natural satellite. With the invention of photography in the early 19th century, photographic records of the Moon were studied. However, just as we do not know everything about the Earth, we know even less about the Moon. Even though *Lunar Orbiters* and the *Apollo* astronauts extensively photographed the areas near to the *Apollo* landing sites, and so the locations of craters and ridges in these areas are well known, there are areas that have not been imaged in great detail, and much of the lunar surface is known only approximately.

Future missions to the Moon include Japan's first lunar explorer the SELenological and ENgineering Explorer (*SELENE*), which was launched in 2006 and will investigate the origins of the Moon, its elemental and mineralogical composition, its geography, its surface and sub-surface structure, the remnant of its magnetic field and its gravity field. Also, a mission to map the Moon is planned by NASA using *Lunar Orbiter Laser Altimeter (LOLA)* This will gather more than four billion measurements of the Moon's surface altitude. Along with high resolution images taken from onboard the spacecraft, the map should offer a very detailed three-dimensional model of the Moon.

The last astronaut to set foot on the Moon was Gene Cernan in December 1972 and, according to NASA's Vision for Space Exploration, astronauts will return to the Moon as early as 2015. On a manned Moon base, astronauts will learn how to live on an alien world before an attempt to another planet, such as Mars, is undertaken. Information on potential future manned Moon landing sites and missions is included in Chapter 11 – "The Future".

Mars

Mars, also known as the Red Planet, is further away from the Sun than the Earth and orbits at about 1.52 Astronomical Units. Although Mars has the second most eccentric orbit of the planets in the solar system, second only to Mercury, the difference between its closest approach to the Sun and its furthest is only about 0.29 Astronomical Units. The atmosphere on Mars is mainly carbon dioxide and it is much less dense than the Earth's. On the surface of Earth the pressure is 1.01 bar, on the surface of Mars it is only about 0.006 bar. To find the same pressure on Earth, you would need to climb to about 35 kilometres high. There is enough activity in Mars' atmosphere to generate winds strong enough to move around grains of sand and dust and so sand dunes form and move, and dust storms can cover the whole planet. Mars is the most hospitable planet in the Solar system for terrestrial type life, after the Earth, and its day is just over 24 Earth hours long, with its year being nearly 687 Earth days long.

Like the Earth, the axis of rotation of Mars is tilted slightly to the ecliptic. Mars is tilted by just over 25° and the Earth is tilted by just over 23°. This gives rise to the seasons. In the winter hemisphere on Mars, polar ice caps grow in size as the carbon dioxide in the atmosphere freezes. Frozen carbon dioxide is also called dry ice, and is used as a stage effect to produce the illusion of fog. In the summer, the polar caps shrink, leaving a permanent cap of mainly water ice. The temperature on the planet can be as low as –150 °C at the winter poles to about 20 °C on the day-lit equator of the planet. In the mid latitudes, the temperature varies daily between about –60 °C at night and 0 °C during the day.

When Mars is closest to the Earth, it can appear as bright as magnitude –2.9, but even when it is furthest away, it will still be a bright object in the sky, at magnitude –1.0. Even with the naked eye, Mars can be seen to be a reddish orange colour. This is mainly due to the iron oxides, better known as rust, in the dust and soil covering the planet. With the aid of a telescope, dark surface features and the brilliant white polar caps can be seen. The surface of Mars is similar to Earth's with impact craters, volcanoes, valleys and deserts. The Martian volcano, Olympus Mons, is the highest mountain in the solar system, standing at about 27 kilometres above the average land height around it. Valles Marineris is the largest canyon, at about 4 kilometres deep and over 4,000 kilometres long. Mars is just over half the width of the Earth and the gravity on the surface is just over a third of that felt on the surface of the Earth.

Mars has two small moons, Phobos and Deimos, which are probably captured asteroids. They are very dim as seen from the Earth, being magnitudes 10.5 and 11.5 at their brightest.

Between 1960 and 1964, six Soviet Union and US missions that were due to visit Mars failed due to various technical difficulties. Eventually, in July 1965 a successful mission was achieved. *Mariner 4* passed within 9,920 kilometres of the planet's surface and returned 22 close-up photographs. The first successful lander was in 1972, when the Soviet Union's *Mars 3* touched down. However, it failed after relaying 20 seconds of video back to the orbiter. The USA's first successful lander and orbiter missions were the *Viking 1* and *2* spacecraft, which both touched down in 1976. Both landers searched for micro-organisms, took panoramic colour photographs and also monitored the Martian weather. The orbiters also mapped the planet's surface, acquiring over 52,000 images. The *Viking 1* lander continued operating until 1982, and the *Viking 2* lander until 1980. The first rover was aboard NASA's *Pathfinder* mission, which arrived in 1997. The six-wheeled rover, called *Sojourner*, was about the size of a microwave oven and investigated Martian rocks and boulders and sent back geological information. The next successful rovers were NASA's *Mars Exploration* Rovers, called *Spirit* and *Opportunity*, which started roaming about Mars in January 2004. Figure 10.2 shows the difference in size between the *Mars Exploration* Rovers and the *Sojourner* rover.

The rovers are about as big as a reasonably sized dining table. They can travel further in a day than *Sojourner* did in its lifetime. By February 2007, *Opportunity* had travelled over ten kilometres. The rovers' prime mission was to last for 90 Martian days, a little over 92 Earth days, but it was thought that the environmental conditions, such as dust storms, might cut the mission shorter. Other environmental factors that could restrict the rovers' life is the elliptical orbit of Mars, which meant that for part of the year the amount of sunlight each rover received would be reduced, and so the amount of power the solar panels could produce would also reduce. Another limit to the rovers' lifetime was expected to be an accumulation of dust on the solar panels, which would slowly reduce the amount of power produced. However, it seems that the

FIGURE 10.2 Members of NASA's Mars Exploration Rovers Assembly, Test and Launch Operations team with One of Their Rovers and a Flight Spare of the Sojourner Rover.
Image courtesy NASA

wind cleans some of the dust off. Mars' elliptical path also means colder nights, and so more electrically powered heating is required to keep the batteries warm enough to work. The rovers have continued operating far longer than expected. Figure 10.3 shows the rover *Opportunity* near the rim of a crater on Mars.

Although the atmosphere on Mars is very thin, it still causes frictional heating as spacecraft pass through it, and so landers need some form of heat shield for protection. However, as the atmosphere is so thin, parachutes, which can be used to slow the rate of descent, cannot slow a lander sufficiently for a soft landing and so other forms of landing system are also required. The *Viking* missions used parachutes and retrorockets for landing, whereas the *Pathfinder* and *Mars Exploration* Rovers used a combination of parachute, airbags and retrorockets. ESA's *Beagle 2* was designed to use a parachute and then airbags to soften the landing, with no retrorockets. Unfortunately, communication was never established with *Beagle 2* after it left its Mother ship, *Mars Express*.

Mars is the most hospitable planet, other than Earth, in the solar system, and so it is likely to be the first planet humans visit.

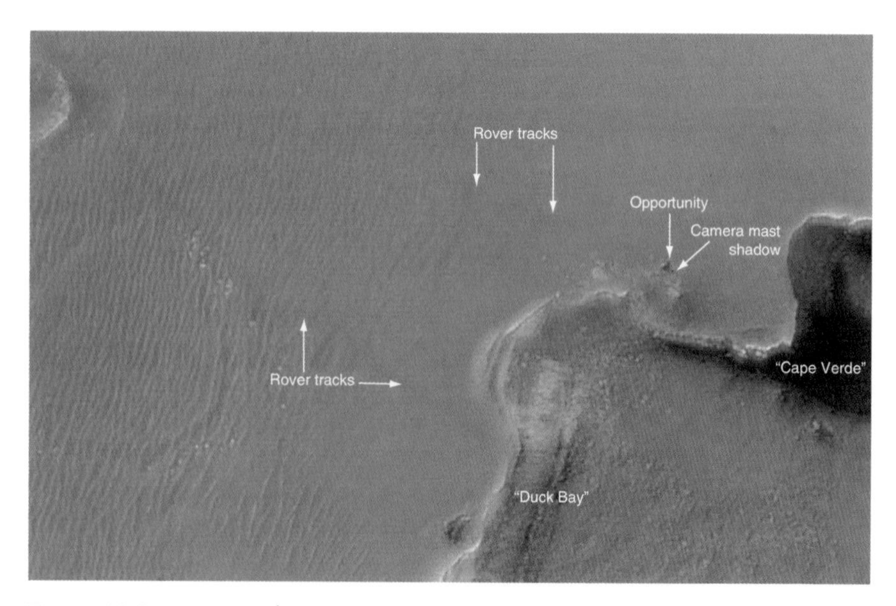

FIGURE 10.3 An Image from NASA's Mars Reconnaissance Orbiter Shows the Mars Exploration Rover Opportunity Near the Rim of "Victoria Crater." Image courtesy NASA/JPL/UA

The flight time to reach and enter orbit around Mars is usually between 200 and 300 days, depending on the amount of propellant carried and the position of the planets at the time of the launch. This would mean a return trip, which requires the correct planetary positions for the forward and return trips, would probably take about 1,000 days. The longest anyone has so far stayed in space was 438 days, this record is held by cosmonaut Valery Polyakov, who stayed on the *Mir* space station in Earth orbit and returned to Earth on March 22, 1995. Quicker journeys require more propellant, other propulsion methods or the use of gravity assist, such as around Venus. However, most solutions carry their own problems. For example, the temperature near Venus is about four times hotter than at Mars, and so heat protection equipment will be required for the first part of the journey, and redundant for the rest of the trip.

The radiation levels outside of the Earth's protective magnetic field will also pose problems to astronauts on a journey to Mars, or anywhere else in our solar system, as described in Chapter 8 – "Humans in Space". More information on the requirements for long duration, outside Earth orbit manned missions is included in Chapter 11 – "The Future".

Jupiter

Jupiter is the fifth planet out from the Sun at about 5.2 Astronomical Unit. It is the largest planet in the solar system, being about 11 times the width of the Earth. It is also the heaviest, being over twice as heavy as all the other planets in the solar system combined. Jupiter is another bright object in the Earth's night sky and can reach up to about magnitude –2.5.

Jupiter is a gas giant and is composed mainly of hydrogen and helium, with a small amount of methane, ammonia, ethane and other chemicals. We cannot penetrate deep into Jupiter's atmosphere using present technology, and so the exact composition of the interior of the planet is unknown. However, theoretical calculations speculate that it probably has a core of solid hydrogen, where the pressure could be a 100 million times greater than on the surface of the Earth, and the temperature could reach 30,000 °C. It is thought that there is probably no clearly defined boundary between the solid core and the sea of liquid hydrogen above it, more of a slushy mixture and maybe a bubbly mixture between the liquid and the gas atmosphere. The "surface" of a gas giant, for the purposes of measuring its radius, is taken to be where the pressure is one bar, or equal to that found on the surface of the Earth. At this surface, the temperature is about –100°C and the gravity is about 2.4 times that on the Earth.

When seen from the Earth through a small telescope or binoculars, Jupiter can be seen as a disk. Through larger telescopes, bands of vivid colour and the Great Red Spot can be seen. The colours include blues, browns, shades of white, reds and yellows, all of which correspond to different depths in the atmosphere. The blues are the deepest and can only be seen through holes in the upper clouds. The browns are the next highest level, followed by the white upper clouds. The red colours, including the Great Red Spot are at the very highest level of the atmosphere. It is thought that the colours are caused by small amounts of chemicals such as sulphur and phosphorous. The Great Red Spot is a vast swirling storm-like feature, although it is not fully understood. It has been observed for over 300 years and varies in size over time. High speed winds blow in opposite directions at different bands of latitude and slight differences in chemical composition and temperature cause the different coloured bands. The dark stripes are known as belts and the lighter areas are known as zones. Other atmospheric storms and interactions can also occasionally be seen, although we can only usually see the clouds at the top of Jupiter's atmosphere, which is a little above the one bar level.

It takes Jupiter a little over 11 Earth years and nine months to complete its orbit around the Sun and as Jupiter's axial tilt is just over 3° only the poles have a significant seasonal change. Jupiter spins very fast about its axis, and at the equator its day is about ten hours long. Different latitudes rotate at slightly different rates. This rate of spin causes the planet to bulge around the equator, and is probably responsible for its very large magnetic field and also the cloud formations. Jupiter's strong magnetic field extends outwards to between a few million and over 650 million kilometres. It has radiation belts similar to the Earth's Van Allen belts, although the radiation trapped within Jupiter's belts is much more intense.

Jupiter has a very faint ring system around it, which was only discovered from images taken by spacecraft passing near the planet.

They are thought to be composed of dust sized grains, which orbit the planet in about five–seven hours. Four of Jupiter's moons, called the Galilean moons, as Galileo was the first to record them in 1610, can also be seen through a relatively small telescope or binoculars. These moons always have the same face towards Jupiter, just as our Moon does towards the Earth. The moons Io and Europa are the innermost objects, and are about the size of the Earth's Moon. They are predominately rocky, although Europa has an icy shell, about 150 kilometres thick. Io has volcanic activity and its surface shows lava flows and other volcanic features. Figure 10.4 shows a fountain of material being ejected from Io.

The two outer Galilean moons are Ganymede and Callisto. These are slightly larger and nearer the size of Mercury and are thought to have a high water or ice content. Ganymede is the largest satellite in the solar system. As of 2006, Jupiter has another 59

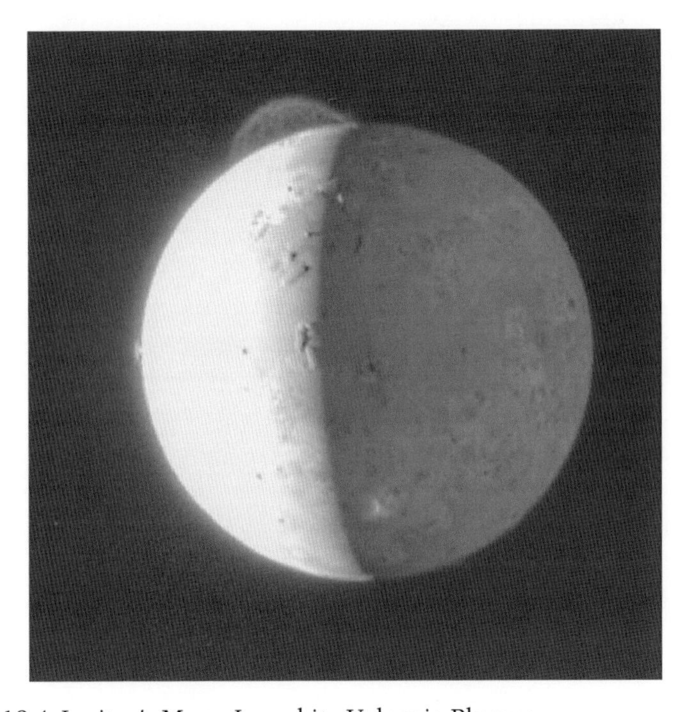

FIGURE 10.4 Jupiter's Moon Io and its Volcanic Plumes.
Image courtesy NASA/Johns Hopkins University Applied Physics Laboratory/Southwest Research Institute

known satellites all of which are smaller than the Galilean satellites, and none of Jupiter's moons has a perceptible atmosphere.

The only mission specifically designed to study Jupiter was NASA's *Galileo* mission. This consisted of an orbiter and an atmospheric probe, which arrived at Jupiter in December 1995 after using Venus and the Earth (twice) as a gravity assist, and then passing close by and imaging the asteroids Gaspra and Ida. *Galileo's* probe measured Jupiter's atmosphere, including its composition and the speed of the wind. The spacecraft *Pioneer 10* and *11*, *Voyager 1* and *2* and also *Cassini* have used Jupiter as a gravity assist and have therefore performed a fly-by past the planet, gathering data as they went.

A manned mission to any of the outer planets is not planned by any space agency. Using current technology, it would take years to complete the journey and the astronauts would need to be shielded from radiation. Landing on a gas giant such as Jupiter would be almost impossible, as the pressures and temperatures would be too intense. However, in the future humans may be able to visit the moons surrounding the outer planets. These could be used as a base to study the planets they are orbiting, and for study of the moons themselves.

Jupiter gives out more heat than it receives from the Sun, which implies an internal heat source. It is thought most of this heat was generated over millions of years as the molecules of gas became more compressed by the planet's gravity and the planet shrank. It is thought this internal heat source causes the atmospheric disturbances on the surface of the planet, unlike the weather on the Earth, which is caused by the heat from the Sun.

Saturn

Saturn is nearly twice as far away from the Sun as Jupiter, at 9.5 Astronomical Unit. It is the second largest planet after Jupiter and about ten times the width of the Earth. It has a very low density, lower than any other planet and lower than the density of water. Saturn is composed mainly of hydrogen and helium, with a small amount of oxygen, iron, neon, nitrogen, sulphur and other chemicals. It is a gas giant like Jupiter, and probably has a similar core, although Saturn's is likely to be proportionally larger than Jupiter's. As Saturn is further from the Sun than Jupiter and also smaller, it is less bright in the night sky, and only reaches about magnitude 0. It is still easily seen with the naked eye and, until William Herschel discovered Uranus in 1781, it was considered to be the furthest planet from the Sun.

At the one bar "surface" the temperature of Saturn is about –130 °C and the gravity is almost the same as on the surface of the Earth. Like all gas giants, Saturn spins very fast about its axis, and at the equator its day is over ten hours long, but at different latitudes it rotates at a slightly different rate. Saturn's equator bulges out and so the planet is flattened almost 30 times more than the Earth is. It also has a large magnetic field, which is probably caused by the rapid spin rate of the planet and the configuration of the core. There are bands in Saturn's atmosphere, similar to Jupiter's although they are more muted and not as spectacular. The colours range from reddish browns to pale orange. The wind speeds in the upper atmosphere are fast, with clouds travelling at 1,500 kilometres per hour, four times faster than on Jupiter. Like Jupiter, Saturn also emits more energy than it receives from the Sun, and so probably has some form of internal heat source, similar in nature to Jupiter's.

From the Earth, Saturn's rings look like solid disks but they are actually composed of billions of small particles of water ice and rock, ranging from the size of a grain of sand to great boulders over one kilometre in diameter, each orbiting the planet. Through a moderate telescope from the Earth, it is possible to see a gap between two of the rings, called the Cassini division. There are at least seven different rings separated by gaps, with the Cassini division being the largest gap. The ring system spreads out over eight times the radius of the planet, but most of the rings are less than one kilometre thick.

Saturn is tilted on its axis a little further than the Earth is, and therefore it has seasons in a similar way to the Earth. As Saturn's rings are aligned to its equator, at Saturn's equinoxes the rings appear to observers on the Earth as a very thin line that is difficult to see. However, near the solstices, the planet is tilted away and therefore the rings appear more open. As it takes Saturn nearly 29.5 Earth years to complete its orbit around the Sun, the change from open to closed rings, as viewed from the Earth, takes about 7.5 years.

Saturn has 56 known moons, most of which are less than 20 kilometres in diameter. The largest, Titan, is larger than the planet Mercury and orbits Saturn in almost 16 days. It is the only satellite in the solar system that has a significant atmosphere, and, like Venus', it is impenetrable by optical telescopes and cameras. The atmosphere is mainly nitrogen, with about 5% of it a mixture of methane, cyanide and other hydrocarbons. On the surface of Titan the pressure is about 1.6 times that of the Earth, which is about the same as you would feel if you dived to a depth of six metres underwater. The temperature is about –178 °C, which is cold enough for the hydrocarbons in the atmosphere, such as methane, to become liquid or even to freeze. It is thought that it may even rain hydrocarbons on Titan. It is also thought that the atmosphere and conditions on Titan are similar to that on prehistoric Earth. Because the atmosphere is so thick and smog-like, the amount of sunlight reaching the surface of the planet is about one thousandth of what reaches the surface of the Earth. However, this is still about 350 times brighter than the moonlight shining from a full Moon onto the Earth.

Saturn's other moons are thought to be ice rich with rocky interiors. Some moons, such as Enceladus show evidence of tectonic activity that has resurfaced the moon, and others, such as Mimas, are covered in impact craters. There are three "shepherding moons" Atlas, Prometheus and Pandora, which are important in keeping the particles in the rings in place.

The first spacecraft to fly by Saturn was *Pioneer 11* in 1979, followed by *Voyager 1* in 1980 and *Voyager 2* in 1981. These missions used Saturn's gravity for a gravity assist, but also provided data about Saturn's magnetic field and atmosphere and returned photographs of Saturn, its ring system and some of its moons.

Cassini Huygens is a joint NASA and ESA mission to Saturn and its moon Titan. NASA's *Cassini* spacecraft took just over seven years to reach the planet, and it entered orbit around it in December 2004. As at May 2007 it was still making an extensive survey of the planet and its moons. ESA's *Huygens* probe left the *Cassini* Mother ship on a 22-day one-way trip to the surface of the moon Titan in January 2005. It was the first probe to land on a world in the outer Solar System. It is hoped that data from *Cassini* and *Huygens* will provide clues about how life began on Earth. Figure 10.5 shows an image taken by the *Huygens* probe on the surface of Titan.

As for all of the gas giants, there are no current plans to send a manned mission to Saturn or its moons.

FIGURE 10.5 ESA's Huygens Probe Recorded This Image of the Surface of Saturn's Moon Titan Image courtesy ESA/ NASA/Descent Imager/ Spectral Radiometer Team (LPL).

Uranus

William Herschel discovered Uranus in 1781, from Bath, England. Herschel was a musician and amateur astronomer who made telescopes to observe the stars. At first he thought Uranus was a nebula or a comet, but after further observations he realised it was a planet. In 1787 Herschel also discovered Uranus' two largest moons, which were later named Titania and Oberon.

Uranus is 19.2 Astronomical Unit from the Sun, which is over twice the distance Saturn is from the Sun. It takes about 84 Earth years to orbit the Sun but its day, at the equator, is only about 18 hours long. At about 70° south its day is nearer 16 hours long.

Uranus is tilted almost horizontally so it has some very strong seasonal variations during its year. During some of the year, the South Pole is pointed directly towards the Sun, during other times, the North Pole is pointed directly to it, and sometimes the Sun is directly over the equator. It is not known what caused the planet to tip over, but it could have been hit by a body about the size of the Earth sometime in its early history, which flipped it over. The International Astronomical Union has defined the north pole of a planet to be the pole that lies north of the ecliptic plane. This means that the north pole of Uranus is tilted 82° and the planet rotates in a retrograde direction. The gravity at the one bar "surface" is slightly less than the Earth's gravity, and the temperature is about –197°C.

Uranus is the third largest planet by diameter, although it has less mass than Neptune. It is a gas giant like Jupiter and Saturn and is about four times the diameter of the Earth. It has an atmosphere of mainly hydrogen and helium with about 2% methane and small amounts of water and ammonia are also present. The methane absorbs red light from the Sun and reflects the other colours to produce the distinctive blue-green colour of the planet. The planet is thought to have a liquid ocean of mainly hydrogen and maybe a very small core of silicate rich material. The winds at the top of the atmosphere reach a speed of several hundred kilometres per hour.

Uranus has rings around it, but these were only discovered by chance in 1977. Astronomers observing stars as Uranus passed in front of them, known as occultations, noticed that the stars disappeared briefly before and after the planet moved in front of them. Further studies showed that there were at least nine thin dark rings around Uranus. There are now known to be at least 11 rings or partial rings around the planet that contain dark particles that range from the size of dust to the size of boulders. Uranus' magnetic field is tilted by 60° to its rotational axis, which is more than for any other planet in the solar system. The Earth's magnetic field is tilted by about 11°. Uranus' magnetic field is not centred at the centre of the planet but is offset by about a third of the diameter of the planet to one side.

Uranus only ever reaches about magnitude 5.5 when viewed from the Earth, and so can only be seen with the naked eye under very favourable, dark sky conditions. Even then it only looks like a faint star. Through a small telescope it can be seen as a small bluish green disk, with no other details visible.

Uranus has five major moons that are between about 500 and 1,600 kilometres in diameter and 22 known moons that are less than 150 kilometres in diameter. The major moons always show the same face to Uranus, just like the Earth's Moon. All five show signs of impact craters and tectonic activity and are thought to be composed of ice and rock. Uranus' moons can only be seen from the Earth with a large telescope or with CCD imaging.

Only one spacecraft has passed near to Uranus, and that was *Voyager 2* in January 1986. *Voyager 2* provided more information on the planet than had been gathered from the Earth in the preceding 200 years. It discovered new rings and sixteen new moons. No more missions to Uranus are currently planned.

Neptune

Because the orbit of Uranus did not follow the same path as theoretical calculations predicted, another planet was suspected to exist outside the orbit of Uranus, which was perturbing its orbit. In 1846, two young mathematicians independently calculated the predicted position of the new planet. They were John Couch Adams of Cambridge, England, and Urbain John Joseph LeVerrier of France. Unfortunately, neither managed to convince astronomers in their own countries to look for the planet, but Johann Gottfried Galle, a German astronomer turned the prime telescope of the Berlin Observatory to look for it and found it on the first night within 1° of the predicted position. Adams and LeVerrier share the credit for discovering the planet. The largest moon, Triton, was discovered 17 days later, but a second moon was not discovered until 1949, when Gerard Kuiper discovered Nereid. Neptune only reaches magnitude 8 and so can only be seen through a telescope or binoculars from the Earth.

At 30 Astronomical Units, Neptune is the furthest gas giant from the Sun and it is over twice the distance Uranus is from the Sun. It takes about 164 Earth years to orbit the Sun but its day, at the equator, is only about 16 hours long. Neptune is the fourth largest planet by diameter, but it has more mass than Uranus. It is just less than four times the width of the Earth. It has an atmosphere of mainly hydrogen and helium with about 1.5% methane. A small amount of water and ammonia are also present. Like on Uranus, the methane absorbs red light from the Sun and reflects the other colours to produce the blue-green colour, but it is not yet known what causes the bright blue clouds that can be seen on it.

Also like Uranus, Neptune is thought to have a liquid ocean of mainly hydrogen and maybe a very small core of rocky material. At the equator the wind carries the clouds over 2,000 kilometres per hour in an east to west direction. The planet spins the same way round as the Earth and it is tilted about 28°, which is a little more than the Earth's tilt, and so it also has seasons. Neptune has a magnetic field, but like Uranus, it is offset from the centre of the planet, and tilted by about 47° from the axis of rotation. It is not known why this is so. The gravity at the one bar "surface" is slightly more than the Earth's gravity, and the temperature is –201 °C.

Neptune also has a ring system, which was first noticed in the mid-1980s by two independent teams of scientist who were watching the occultation of stars behind the planet. These were thought to be only segments of rings although they have now been shown to be complete rings with faintly visible material connecting brighter clumps of dust.

Neptune's largest moon, Triton, has a thin atmosphere of nitrogen and is geologically active. In clear conditions, Triton can be seen from the Earth through a medium sized good quality telescope. As Triton's orbit around Neptune is retrograde, it is not thought that it could have formed near Neptune, but it was probably captured later. Neptune's other moons are much smaller and range from about 50 to 400 kilometres in diameter.

Voyager 2 is the only spacecraft to have passed near to Neptune, when it performed a fly-by in August 1989. The images it sent back showed the cloud structure and motion as well as rings and discovered six of the named eight moons. No new missions to Neptune are currently planned but the planet is still studied using the *Hubble Space Telescope* and Earth-based telescopes and equipment.

Dwarf Planets

In August 2006, the International Astronomical Union (IAU) General Assembly met in Prague and agreed on the new definition of a planet. A new distinct class of objects called dwarf planets was also decided. Pluto, which had been classified as a planet since Clyde Tombaugh discovered it in February 1930, became a dwarf planet, and the number of planets in our solar system was reduced from nine to eight.

Also placed in the dwarf planet category is Ceres, previously the largest known asteroid, and the temporarily named object 2003 UB313, also nicknamed Xena, which has since been named Eris. There are other potential objects that may become labelled as dwarf planets, when more information is gathered on them. There have been no spacecraft that have passed near to any of the dwarf planets yet, although some are on their way.

Pluto

Pluto is 39.5 Astronomical Units from the Sun, and it takes about 248 Earth years for it to orbit. Its day is a little over six Earth days and its rotation is retrograde. Its orbit is highly eccentric, and sometimes it comes closer to the Sun than Neptune, as it was between January 1979 and February 1999. Its orbit is also very inclined. Like Uranus, the axis of rotation is almost at right angles to the plane of its orbit, and so the apparent path of the Sun across the sky is very irregular. The temperature on Pluto is on average about –223 °C.

Pluto is about 2,400 kilometres in diameter, which is about a fifth of the diameter of the Earth, and it has three moons, called Charon, Nix and Hydra. Charon is the largest, being about half the diameter of Pluto, whereas Nix and Hydra are thought to be between 60 and 200 kilometres in diameter. Pluto and Charon are synchronously locked, and so keep the same face towards each other all of the time. Because of this and their relative sizes, the Pluto-Charon system is therefore considered by many scientists to be a binary or double planet system.

It is thought that Pluto and Charon are composed of a silicate rock core, with water and methane ice surrounding it, similar to Neptune's moon Triton. However, Pluto's surface contains nitrogen ice and some methane and Charon's surface is mostly water ice. Pluto seems to have a very tenuous atmosphere, probably composed of nitrogen with some methane. The gravity on the surface of Pluto is about 1/25th that of the surface of the Earth.

Seen from the Earth, Pluto is never brighter than magnitude 14, and so can only be seen through a moderately sized or large telescope. However, it will only appear as a pinpoint of light, with no details observable. Because it is so far away, even the best images from the *Hubble Space Telescope* show Pluto and Charon as a fuzzy blob.

In January 2006, NASA's *New Horizons* spacecraft was launched. This should fly by Pluto and Charon in 2015, and it will then continue through the Kuiper Belt and eventually leave the solar system.

Ceres

In 1801 the Sicilian Giuseppe Piazzi discovered a bright object that was not on any star maps. At first he thought it was a comet, but after further observations, it appeared more like a small planet. The object was named Ceres. Within a few years three more small bodies were discovered and by the end of the 19th century, several hundred were known. William Herschel proposed the name asteroid, which is Greek for "star like", for this new type of object.

Ceres is about 2.7 Astronomical Units from the Sun, so it is between Mars and Jupiter, in what is known as the asteroid belt. It is almost spherical, and has a diameter of about 950 kilometres. It is the largest object in the asteroid belt, and is about a third of the mass of the total of everything in the belt. It rotates in just over nine hours, and orbits the Sun in a little over four and a half Earth years. From the Earth, Ceres appears about magnitude 7 and is just possible to see with the naked eye.

Asteroids are classified by their position in the solar system and their infrared composition. Ceres is a C-type, which means it is a carbonaceous asteroid that is rich in carbon and other complex organic compounds with some chemically bound water.

In 2007, NASA will launch the *Dawn* spacecraft on a mission to study Ceres and also the asteroid Vesta. It is expected to reach Ceres in 2015.

Eris (2003 UB313)

In July 2005, an object larger than Pluto was discovered orbiting at about 97 Astronomical Units from the Sun. This is over double the distance Pluto is from the Sun. The object, now called Eris, is currently the largest dwarf planet and has a highly elliptical orbit, so that its distances from the Sun ranges from about 38 to 97 Astronomical Units over the course of its almost 557 Earth year orbit. Mike Brown and his team used the Palomar Observatory's Samuel Oschin telescope to discover it. It is currently about magnitude 20. It has a diameter of about 3,000 kilometres, which is over 700 kilometres larger than Pluto. Eris has a small moon called Dysnomia. There are no plans yet for any spacecraft to visit Eris.

Small Solar System Bodies

At the International Astronomical Union (IAU) General Assembly 2006, it was also decided that in the solar system there are planets, dwarf planets, and that:

> All other objects except satellites orbiting the Sun shall be referred to collectively as Small Solar-System Bodies.

This includes most of the solar system asteroids, most Trans-Neptunian Objects (TNOs), which are small icy bodies outside of the orbit of Neptune, comets, and other small bodies.

Asteroids

Asteroids are small objects in the solar system that are too small to be planets or dwarf planets. They are classified by their position in the solar system and by their type, which is related to their composition, usually rocky or metallic. There are different classification systems, but the types usually used are based on the Tholen taxonomy, which was produced by the American astronomer David Tholen in 1984. As more information is gathered about asteroids using more advanced equipment, further classes and sub classes are used to describe each asteroid. The Tholen classification is based on the albedo and also the colour of the asteroid. From this information, their composition can be inferred. Most asteroids fit into three main classifications of C-type, S-type and M-type. The C-types are dark carbonaceous objects, which include asteroids that contain carbon, organic compounds, silicates and other minerals that contain water. They are extremely dark. The S-types are silicaceous, that is, stony or rocky. They consist of metallic iron mixed with iron silicates and magnesium silicates. These are relatively bright asteroids. The M-types are metallic and most consist mainly of metallic iron and are relatively bright. As more is discovered about individual asteroids and comets, the differentiation between the two is becoming less clear, and some asteroids have been reclassified as comets.

Most asteroids are between the orbits of Mars and Jupiter in the main asteroid belt, and are called main belt asteroids. This covers the distance from the Sun from about 2 to 4 Astronomical Units. It is estimated that there are over a million objects in the belt that are over one kilometre across, with about 1,000 objects over 30 kilometres long, and about 26 known to be larger than 200 kilometres across. The brightest is Vesta, which occasionally reaches magnitude 6. Asteroids that come within 1.3 Astronomical Units of the Sun are known as Near Earth Asteroids or NEAs. Trojans are asteroids located in the Lagrangian points 4 and 5 of some planets. The majority are in Jupiter's Lagrangian points, 60° ahead and behind the planet on its orbit around the Sun. There are also a few asteroids in orbit at the Lagrangian points of Mars and Neptune. Asteroids between the orbits of Jupiter and Neptune are called Centaurs, although these are probably more like comets in composition than asteroids. Asteroids further out are called Trans-Neptunian Objects (TNO's) and most orbit at a distance from the Sun comparable with Pluto.

It is thought that many asteroids have other asteroids orbiting around them, called asteroid moons, satellites or companions. Asteroid Ida, for example, has a moon called Dactyl. If the two asteroids are of a comparative size, they may be considered asteroid binaries or an asteroid double system. It is possible for an asteroid to have more than one moon.

The first fly-by of an asteroid was by the spacecraft *Galileo* in October 1991, when it passed near to asteroid Gaspra. The S-type asteroid is irregularly shaped and has a cratered surface, as can be seen in Figure 10.6.

In August 1993 the *Galileo* spacecraft passed closer to the asteroid Ida than it did to Gaspra, and so the surface was imaged in greater detail. Ida's moon Dactyl was also discovered at that time. The *Near Earth Asteroid Rendezvous (NEAR/Shoemaker)* mission investigated the composition of asteroids in a near Earth orbit. After completing its mission in February 2001, it made a controlled

FIGURE 10.6 The Asteroid Gaspra.
Image courtesy NASA/JPL

descent to the asteroid Eros, where it continued to send data for a while. In 2011, NASA's *Dawn* mission should reach an orbit around asteroid Vesta, and then in 2015 it should orbit asteroid Ceres. These two asteroids are very different as it is thought that Vesta has no or very little water but Ceres may have a 100 kilometres layer of water ice or liquid water beneath its crust. By discovering more about the composition of these asteroids, planetary formation may be better understood.

Comets

The first recorded sighting of a comet was by Chinese astronomers in about 2300 BC. Comets were discovered to be in orbit around the Sun and not a phenomenon in the Earth's atmosphere by Tycho Brahe, the Danish astronomer, in 1577, as he systematically studied them. In 1758, the British astronomer Edmund Halley showed that comets have highly elliptical paths and from his work and previous records of comet appearances, he predicted the return of a comet. That comet was later named comet Halley, and is probably the best known comet. In 1950, the American astronomer Fred Whipple proposed that comets were made of water ice with dust and dirt in them. Comets are now often called dirty snowballs or icy mudballs. Comets are mainly frozen water but also contain small amounts of methane, carbon monoxide and carbon dioxide and other carbon containing molecules and also silicates.

At the centre of a comet is the nucleus, which is solid and is thought to be around 20 kilometres or less in diameter. When the comet comes into the inner solar system, the heat from the Sun makes some of the ice change straight from a solid to a gas. This is known as sublimation. The gas and any solid dust grains that had been trapped are released and form a coma around the nucleus. The coma can expand to a diameter of over 1.5 million kilometres. As the gravity of the nucleus is so weak, the radiation pressure from the sunlight can strip the particles in the coma away and create a dust tail in the direction opposite to the Sun. The particles in the tail reflect sunlight, and so it can sometimes be seen from the Earth. The tail can stretch for tens of millions of kilometres. The solar wind also affects the coma. Some of the gases in the coma are converted into ions by the high-speed electrically charged particles flowing from the Sun in the solar wind. These ions stream away from the coma, leaving an ion tail. The ion tail can be several hundred millions of kilometres long, and can sometimes be seen as they glow as the absorbed energy from the Sun is released.

A comet is classified by how long it takes to orbit the Sun. If it takes less than 200 years, it is classified as a short period comet, and if takes over 200 years it is a long period comet. Each time a comet passes near to the Sun, it loses some ice and gas and, after all the ice and gas is lost, a rocky object, very much like an asteroid, is left. It is thought that perhaps half of the Near Earth Asteroids were once comets.

The first spacecraft to fly-by a comet was the *International Comet Explorer* (*ICE*), which was launched in August 1978 and passed comet Giacobini-Zinner in September 1985. Five missions from different countries were flown to comet Halley in 1986. In 2004, the NASA's *Stardust* mission gathered samples from the coma of comet Wild 2 and returned them to the Earth and in 2005 NASA's *Deep Impact* created a deep crater on comet Tempel 1 and therefore the pristine interior of the comet could be investigated.

As mentioned previously, ESA's *Rosetta* spacecraft is due to go into orbit around comet 67P Churyumov-Gerasimenko in 2014 and release a small probe that will land on the comet's nucleus and hopefully provide more information about the nature of comets.

11. The Future

Space exploration is an expensive challenge and a single nation can only investigate a small portion on its own. This is why, in 2007, 14 space agencies around the world developed *The Global Exploration Strategy: The Framework for Coordination*. This strategy provides a vision for robotic and human space exploration and focuses on destinations within the solar system where humans may, one day, live and work. It is not a proposal for a single global program, but instead allows nations to collaborate and strengthen both individual and collective projects. This should mean that technology and knowledge is transferred between space agencies and that time and effort is not wasted repeating similar projects.

The next 20 years or so will therefore see a more global approach to space exploration, with many nations contributing to the scientific, technical and financial challenges posed. Human missions to the Moon will be used to test systems, operations that could lead to sustainable human missions to Mars and beyond, and also determine whether we can live on other worlds. Robotic missions will explore our solar system and may provide answers about the origins of the planets and whether life exists beyond Earth. Space technology has already produced spin-offs for use on the Earth. For example, robotic instrumentation developed to search for life on Mars is now being adapted into a portable tuberculosis diagnostic machine for use in the developing world. It is hoped that other spin-offs will be developed, probably in the fields of medicine, agriculture and environmental management, which will benefit all mankind.

Each nation has a plan for the future that suits their scientific and political agendas. In 2004 the president of the USA, George W. Bush, presented a vision for the future of US Space Exploration. The vision included a robust space exploration program, with four major goals:

1. Implement a sustained and affordable human and robotic program to explore the solar system and beyond
2. Extend human presence across the solar system, starting with a human return to the Moon by the year 2020, in

preparation for human exploration of Mars and other destinations
3. Develop the innovative technologies, knowledge and infrastructures both to explore and to support decisions about the destinations for human exploration
4. Promote international and commercial participation in exploration to further US scientific, security and economic interests

Some of the challenges that must be faced in order to meet these goals include efficient power generation and energy storage, health care for human explorers, autonomous operation and smart decision-making for robotic explorers, planetary resource extraction and utilization, on-orbit spacecraft servicing and the building of safe habitats with efficient life support and environmental controls. In order to support human and other life far from Earth, resources will need to be conserved and as much as possible must be recycled, and so the systems for this must be improved. Communication and navigation facilities that can be used outside of Earth's orbit also need to be developed further, as does transportation in space and on the surface of the Moon or other celestial object and also human and robot cooperation. Minimum mass requirements and the need for reliable operation in a high-radiation environment constrain these already complicated challenges.

Commercial Space Flight

Because of the Outer Space Treaty, described in Chapter 1 – "Introduction", nations cannot claim ownership of outer space, including the Moon. When the *Apollo 11* mission landed on the Moon in 1969, the US astronauts Neil Armstrong and Buzz Aldrin planted the US flag as a sign of pride, but not of sovereignty. They also left a plaque on the Moon that said:

> HERE MEN FROM THE PLANET EARTH FIRST SET FOOT UPON THE MOON JULY 1969, A.D. WE CAME IN PEACE FOR ALL MANKIND.

No country can support or enforce a private individual's claim to ownership of the Moon or other celestial body without breaching the treaty. Therefore, any certificate of ownership of an area of the Moon, a star or any other celestial body is unenforceable and does not confer title to the area or object.

There is some uncertainty to the legal status of exploiting the natural resources on the Moon or other celestial bodies. Only nine countries have ratified a separate treaty, called the Moon Treaty. One of the principles of this treaty is the commitment that all nations will share in the management and benefits from exploiting these resources. However, benefits could be shared without a contribution to the investment involved in gaining the benefit, and so spacefaring nations, such as the USA, have not joined the treaty. However, if the USA did allow anyone a licence to mine the Moon or other celestial body, many nations would resent the USA "stealing" from them, and so it is expected that the legal situation would need to be confirmed before any such activity is begun.

As long as the Outer Space Treaty is not breached, commercial companies can have access to space. In 1965, NASA launched the first commercial communications satellite, *Early Bird (INTELSAT 1)*. This was capable of forwarding television broadcasts between Europe and the USA. The Communications Satellite Corporation (COMSAT) operated the satellite in cooperation with the International Telecommunications Satellite Consortium (INTELSAT). Before the 1980s, all commercial satellites were launched by the USA on vehicles owned by the government, including the Space Shuttle. In 1980, the first commercial space transportation company, Arianespace, was created. Arianespace initially used the *Ariane* family of launch vehicles, but now also uses the *Soyuz* launch vehicles. After the *Challenger* Space Shuttle disaster in 1986, commercial payloads were banned from flying aboard the Space Shuttle. This prompted more commercial space launch companies to be developed. By 2002, more than US$95 billion of economic activity was due to US commercial space transportation and related services and industries. For the last few years about a quarter of all the rocket launches into space have been by commercial companies. All commercial launches currently use expendable launch vehicles, which are used only once. This means the price for getting anything into space is high. The development of reusable launch vehicles would reduce the price considerably, and work is being conducted in many areas to do this.

Space Tourism

The high price of getting anything into space has meant that the opportunity for humans to enjoy space travel has been limited to government-employed astronauts and a few fee-paying individuals, who have paid around US$20 million for the experience. For their money, these private astronauts got a return trip in a *Soyuz* rocket and time aboard the International Space Station. As all of these travellers have assisted with research on the Station, they can be thought of more as explorers or independent researchers, rather than tourists. It will probably be at least a few more decades before tourists can stay in space hotels or visit the Moon, but the first step towards such trips is being developed by companies who aim to provide sub-orbital flights.

In order to encourage private companies to help to open up the space frontier, particularly human spaceflight, the X-Prize was formed. This was modelled on the US$25,000 Orteig Prize, which was offered to the first pilot to fly non-stop between New York and Paris. In 1927, an unknown airmail pilot won the prize. His name was Charles Lindbergh. The US$10 million X-Prize, sponsored by the Ansari family, was awarded to the first private team to build and launch a spacecraft capable of carrying three people to 100 kilometres above the Earth's surface, twice within two weeks. Between the teams, over ten times the amount of the prize purse was spent trying to win the competition. Mojave Aerospace Ventures won the prize on October 4, 2004, for the flight of *SpaceShipOne*. Although *SpaceShipOne* was capable of carrying three people, only the civilian test pilot, Mike Melvill, was onboard for the flights. The X-Prize brought the idea of personal spaceflights and space tourism to the attention of the public, and due to the development of reusable launch vehicles and associated equipment, over the next few years relatively cheap, but short, sub-orbital flights should be available to the general public. The next stage of development will be to provide low Earth orbit flights.

The sub-orbital flights being planned will allow tourists to experience the thrill of rocket-powered acceleration. This will take them to over three times the speed of sound. They will see the sky turn from blue to mauve and then to black, and the curve of the Earth will be visible, with its thin blue band of atmosphere surrounding it. Most flights will provide a few minutes of weightlessness, when tourists will be able to leave their seats. After the event, they will be awarded their astronaut wings.

All space vehicles will be licensed by national or regional safety organizations, such as the FAA in the US and the European Aviation Safety Agency. As the development of space planes and other means of space tourism is new and as there are various technical solutions to many of the challenges, the best designs in terms of safety and reliability have yet to be determined. Therefore, at least for the FAA, there are minimum standards and specific requirements for safety of operation that the sub-orbital spacecraft, as well as the spaceport operations and adjacent facilities, must meet. Until at least 2010, all vehicles and spaceports will be inspected and licensed on a case-by-case basis.

Different companies are developing different launch mechanisms for their tourist flights. One group has a design that involves a reusable helium balloon, which will lift a spacecraft to an altitude of over 20 kilometres, where the spacecraft's rocket engines will fire and propel the crew to over 100 kilometres. The major disadvantage of this type of system is that it requires very calm or no wind conditions, before it can lift safely. The rocket will be capable of performing several missions, after it has returned to Earth by parachute. Other companies are investigating the more conventional vertical takeoff rocket launch and vertical landing, such as used by the *Gemini* and *Apollo* missions, and others are developing vertical takeoff and horizontal landing vehicles, similar to NASA's Space Shuttle. A horizontal takeoff and horizontal landing design, such as used by *SpaceShipOne*, which won the X-Prize, is popular as most designs do not require a specialised spaceport, and could take off and land at conventional airports.

Some companies are looking towards orbital flights and beyond. One company is designing a trip that will use airships and ion engines to provide access to a low Earth orbit. Flying an airship all the way from the ground to orbit in one stage is not practical. The winds near the surface of the Earth would damage an airship large enough to reach orbit, and an airship that could fly from the ground to the upper atmosphere would not be light enough to reach space. Therefore a three-part system is planned. The first part is an atmospheric airship, which will travel from the surface of the Earth to over 42 kilometres high. The vehicle would be crewed, and would also be capable of carrying passengers or cargo. The second part is a sub-orbital space station, called a *Dark Sky Station*, which would probably be held aloft by an airship, and would act as a waypoint to space. The passengers or cargo would be transferred at this point from the atmospheric airship to an orbital airship. Although the *Dark Sky Station* is currently being developed, whether this size of vehicle is feasible, either from an engineering or economic standpoint, is debateable. If the *Dark Sky Station* is a success, however, the orbital airship will leave it and fly directly to orbit, using the buoyancy of a very large craft to take advantage of the few molecules of gas at these altitudes to climb to over 60 kilometres. The designers claim that after this height, it will use electric propulsion to slowly accelerate and therefore climb to a higher orbit and after a few days it will reach orbital velocity. However, it is disputed that the electric propulsion system proposed would be able to overcome atmospheric drag and therefore the spacecraft would not reach the required velocity.

Other companies are developing commercial space habitats for tourists or scientific researchers. Bigelow Aerospace has successfully launched into space two experimental models of a human-habitable module. The models, *Genesis I* and *Genesis II*, are about a third the size of the intended final module. All of their spacecraft are designed to enter space and then expand to increase their volume. The flexible shell of the spacecraft is about 15 centimetres thick and made up of many layers of material, similar to the way spacesuits are made. After the first six months in orbit, *Genesis I* was still in very good shape. The expandable shell and pressurized structure was perfectly intact and no discoloration, due to exposure to high ultraviolet radiation, was detected.

As the number of space tourism providers increases, the costs should reduce. If the demand is great, more spaceports and their supporting infrastructure, including the training facilities, will be required, and a new industry will have been developed.

There are three potential major problems for the space tourism industry. First are the environmental concerns. These include the amount and type of emissions from the launch vehicles, such as carbon dioxide and other, maybe more toxic, chemicals, the huge amounts of energy required to propel the launch vehicle into space and also the impact on the Earth's ozone layer due to the high volume of flights through the atmosphere. Limits imposed on carbon dioxide emissions may make space tourism unfeasible, but if no limits are imposed, the increased volume of carbon dioxide in the atmosphere, due to regular launches, could adversely impact the world's climate. The second potential problem is orbital space debris, as described in Chapter 1– "Introduction", which could cause catastrophic damage to the space vehicles. The third potential problem is the possibility of deployment of weapons in space or on the Moon. There are treaties that currently prevent this last possibility, but the USA has implied that it would not be bound by them if its national security were threatened by attacks on critical space assets.

Future Propulsion Systems

There are many forms of propulsion system that have been proposed but have not yet been used, either because of technological constraints or because they provide no great advantage over current methods. There are other novel forms of propulsion that have already been developed and have even been used to provide propulsion or attitude correction for spacecraft. However, these technologies are still in the development phase and are not yet used for the primary propulsion of spacecraft. This section describes many of the forms of propulsion systems that have been suggested and that are theoretically feasible, although some will probably not be practical for many years.

Solar Sailing

As has been mentioned previously, sunlight consists of tiny packets of light, called photons. When they move, these photons have momentum that can be transferred to other objects, this is called solar radiation pressure. The Sun also radiates a constant stream of particles called the solar wind, which has an associated magnetic field. The most obvious example of the power of these emissions can be seen in the two tails of a comet. As a comet is warmed by the Sun, the surface, which is usually dirty ice, starts to melt, and gas and dust particles are sloughed off to form the comet's tail. The tail is buffeted by both the solar wind and also by solar radiation pressure. The majority of the solar wind particles are protons and electrons, which interact with the gas particles in the tail and form charged particles of gas, called ions. A force, generated by the magnetic field in the solar wind, then pushes this tail of ions outwards, and the ion tail is always seen pointing directly away from the Sun. Each dust particle in the tail is in orbit around the Sun and is pushed away from the Sun by radiation pressure. As the orbital period further away from the Sun is greater than it is close to the Sun, the dust tail curves slightly in the direction the comet came from.

Although the solar wind sounds like it should be easy to harness, in the manner of sailing ships on the Earth's oceans, it is actually the solar radiation pressure that has the most effect on solar sails.

A solar sail, also known as a light sail, can be used to harness the solar radiation pressure from the Sun. However, if it just absorbed the radiation pressure it would be pushed away from the Sun, in a similar way to a comet's dust tail, by the incident momentum of the photons. This would only allow the sail to travel further and further away from the Sun, with no means of controlling the direction of travel. However, if the sail is able to reflect the photons, the amount of momentum gained is doubled. This principle can be demonstrated with the use of a skateboard and a heavy ball. If you stand on the skateboard and someone throws the ball to you (and you catch it), the skateboard will start to roll away, as the momentum of the ball has been transferred to the board. Now if, after catching the ball, you throw it back, the board will move even faster, due to the reaction of the departing ball. The combination of the incident momentum and the reaction from the reflection will push the sail at right angles to its surface. Therefore by changing the tilt of the sail, the direction of the thrust can be controlled and the sail can be steered. It is even possible to slow the spacecraft and therefore direct the sail on a spiral in towards the Sun.

The pressure from sunlight is small. A sail located at 1 Astronomical Unit from the Sun, could, at best, only receive a force of nine newtons per square kilometre of sail. On the surface of the Earth, it takes a little over nine newtons to lift a bag of sugar or a litre of milk. Therefore, to provide enough force to propel a payload attached to a solar sail, the sail must be very large. In the 1970s a NASA team investigated using a solar sail-powered spacecraft to rendezvous with Halley's comet but, due to technical and financial constraints, the mission did not happen. A more recent solar sail test attempt, the Planetary Society's *Cosmos 1*, was to use a much smaller sail than the Halley's comet mission. If the launch rocket had not failed in June 2005 and it had managed to spread its sails, it would have had a total surface area of $600 m^2$, a little bit smaller than a football pitch.

Some spacecraft, such as the *Pioneer 10* mission, have used solar radiation pressure for minor attitude changes and the deployment of solar sails and thin reflectors have been tested in space, for example, by the Russian *Znamya* in 1993, which was designed to reflect sunlight to Russian towns during the dark winter months. NASA also tried to use a solar sail for its *Inflatable Antenna Experiment* in 1996. If it had deployed successfully, it would have been used as a radio frequency reflector. As of June 2007, there have been no spacecraft accelerated by solar sails.

Solar Sail Design

Solar sails can only begin to work when they are outside of the Earth's atmosphere, as within it, the air pressure and wind overwhelm the radiation pressure from the Sun. The sails must therefore be launched aboard a conventional rocket before deploying in orbit. As most rockets provide a relatively small payload area, the sail must be robust enough to be packaged tightly and, as the sails need to be so large and the thrust from the Sun is so small, the sail itself must be as light as possible. Ideally, a thin film of reflective material, such as aluminium, would be used, but this would easily tear during the packing and subsequent deployment. A plastic backing attached to the reflector is therefore required. When in use, the sail must be kept flat, which can be achieved with the aid of structural supports, such as spars similar to those used in a child's kite, or by spinning the sail to keep it stretched out, like a figure skater's skirt when she spins around.

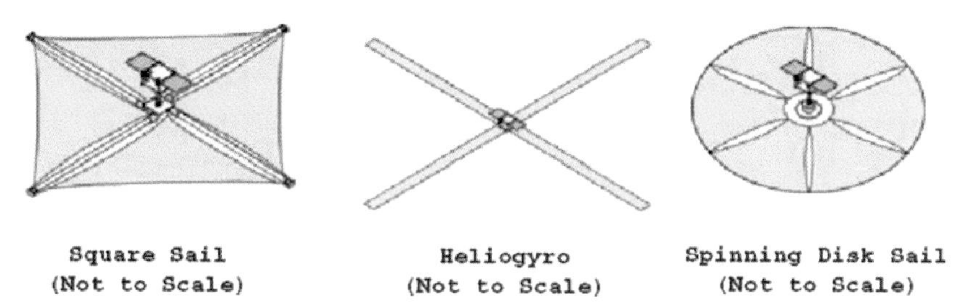

FIGURE 11.1 Solar Sails.
Image courtesy NASA

There are three major configurations for solar sails, as shown in Figure 11.1. These are the square sail, a heliogyro and a disk sail. The square sail has four spars cantilevered from the central hub. The tip vanes can be used to control the attitude of the sail. The major disadvantage with this design is that it is complicated to pack and deploy, and therefore there are more chances of something going wrong.

A heliogyro has the sail divided into long thin strips or blades that are attached around a central hub. The heliogyro spins to maintain a flat surface. By changing the pitch of the blades, the heliogyro can be manoeuvred. Although this design is very simple to pack and deploy, as it just needs to be rolled up and unrolled, the edges of each blade may require stiffeners, adding a complication to the sails.

A disk sail is similar to the heliogyro but it provides one continuous reflective surface. This again uses spin to keep it flat, but it may also require radial spars to keep it flat during attitude manoeuvres and a hoop around the exterior to provide tension at the edges. A further advantage of this design is that the elastic energy stored in the spars during the packing could be used to drive the deployment and unfurl the sail. The result of stored elastic energy can be seen when an arrow is released from a bow, or a jack-in-the-box is released from its box.

Even with very large sails, a solar sail spacecraft will accelerate very slowly, compared to a conventional rocket. However, as long as the sail receives solar radiation, it will constantly accelerate, unlike a rocket that only burns for a short time and then coasts for most of the remainder of its mission, with occasional small thrusts to correct its trajectory and attitude.

Because rocket-driven spacecraft spend most of their time coasting, their orbits are mainly determined by gravity and are described as Keplerian orbits. A spacecraft that can continuously accelerate can travel in non-Keplerian orbits, and so may, for example, be able to circle the Sun entirely above or below the ecliptic or to circle the Earth entirely above or below the equatorial plane. This could allow continuous monitoring of the poles of both the Sun and the Earth, and could also allow continuous communication at high latitudes, without the need for multiple satellites. To do this, the solar radiation force must be similar in magnitude to the local force of gravity, so, with an appropriate sail, it should be possible to hover anywhere in the solar system. By using a sail with the correct characteristic acceleration, the Lagrange balance points of any system could be artificially moved, giving scientific missions more scope to position their instruments and therefore further their investigations. These non-Keplerian orbits are not just restricted to solar sails, but as propellant-driven spacecraft are limited by the amount of propellant they carry with them, they are not usually used in this way. However, for missions of less than about a year's duration, ion drives, described in Chapter 5 – "Propulsion Systems", may be more suitable than solar sails.

As the solar radiation pressure is largest closer to the Sun, smaller sails could be used for inner solar system missions than would be required for outer solar system missions. Solar sail delivery of payloads to Mercury or near to the Sun can be much cheaper than for traditional ballistic trajectories, even with the help of a gravity assist. This is because a smaller launch rocket can be used, as a solar sail can start working from a low orbit and also because it does not have to carry propellant.

Travelling to Mars by solar sail will take longer than by ballistic methods, but it is not restrained by the positioning of the planets in their orbits around the Sun, and therefore does not have restrictive launch windows. One-way missions are not the most effective use of solar sails, as once they have reached their destination, they are still able to continue harvesting the solar radiation and travel further. This makes return missions from planets, asteroids or comets much more feasible than for ballistic missions, which must carry the fuel for the return journey with them. It is possible that one sail could do many journeys between the Earth and Mars, delivering objects or goods that are not time dependent. Other missions include geological surveys, travelling from one asteroid to the next, before returning samples to the Earth and setting off again.

As the solar radiation pressure reduces with distance from the Sun, missions to reach the outer solar system must first be accelerated by looping though the inner solar system to gain speed. When they reach their destination they must then be slowed, in the same way that ballistic transfer missions must be slowed, by either using propellant or by aerobraking. However, if a mission is required to escape the solar system, the slowing mechanism is not required and even though the radiation pressure is low, there is very little friction to slow the solar sail and it can continue at very high cruise speeds. As the solar sail travels away from the Sun, electricity generation from photovoltaic or solar cells would no longer be sufficient and so an alternative form of power would be required to supply the instruments, navigational and communication equipment and also to control the sail. For outer solar system or deep space missions nuclear fission or radioisotope-powered propulsion drives are more suitable, as they generate their own electricity.

There are potential problems with solar sails, including wrinkles, meteorite punctures and electrical charging. If the sail material wrinkles, there may be problems of local hotspots, which could lead to the sail melting in places. Punctures from meteoroids could rip the sail, which may propagate and tear the sail apart. This could be prevented by the introduction of ripstops produced by doubling the sail material, introducing seams or by reinforcing the material, but these will all add to the mass of the sail. Static electricity could cause shorting or arcing across the sail, which may rip the sail. This could be prevented by the addition of electrical shorting devices, although these would also increase the mass of the sail.

Beam Sailing

Instead of just relying on the Sun to power sails, some other form of energy, such as light or even pellets, could be beamed to the sail, either from the Earth or from an orbit in space, and bounced off to produce a thrust. The heavy part of the spacecraft, consisting of the propellant, the energy source and the motor or engine, could then be separate from the spacecraft and does not need to be accelerated, and so it could remain on the Earth or in an orbit around the Earth, Moon or Sun. It could also be used time and again. The major disadvantage of a beamed system is the amount of power required. Another obstacle to a beamed system could be the transmitter. Although the beam should only be used to power spacecraft, it could also cause damage to electronic systems and also scramble radio signals or other transmissions in space, particularly if it were situated in an Earth or solar orbit. There could therefore be political objections to such a system by the nations who are not in control of it.

Lasers

A beam from a laser could be used to propel a vehicle in space. There are a few ways the laser radiation could be harnessed. First, it could be used for photon propulsion, with the photons from the laser beam accelerating a sail in the same way photons from the Sun could propel a solar sail. It could also be used to heat the sail, so that the surface decomposes, known as thermal ablation or just ablation. The gases given off would propel the spacecraft. The laser could also be focused onto a boiler full of propellant onboard the spacecraft. As the propellant expands, it would escape through a nozzle and provide the propulsion.

Photon Propulsion by Laser

A beam from a powerful laser focused onto a sail could accelerate it so that it leaves the solar system faster than a Sun-powered solar sail would. However, the transfer of energy from a laser to a sail is highly inefficient. A gigawatt of laser power would only produce 6.7 newtons of force. Even if this huge power requirement were generated by the Sun, the laser would still need to be focused, probably using a large aperture lens, to direct the energy in parallel beams towards the spacecraft. As a laser is distorted when it travels through the atmosphere, both the laser and the lens would have to be in space. The size and type of laser needed for a photon laser sail has not yet been developed, and there are many technical engineering difficulties involved in building and controlling the lens.

It is estimated that with a ten gigawatt laser and a one kilometre diameter lens, a 30 kilograms sail with a diameter of 195 metres could reach 100 Astronomical Units in about 9.5 months. It has taken NASA's *Voyager 1* nearly 30 years to reach this far with the use of chemical propulsion from the launch vehicle and gravity assists from Jupiter and Saturn. There is a limit to how far a laser beam would be effective before it disperses too much and would no longer propel the sail. At this stage the spacecraft will just coast, continuing its journey outwards. If the solar sail spacecraft above continued to coast, it would reach 1,000 Astronomical Units in a little over eight years. Light from the Sun takes less than six days, or six light days, to reach 1,000 Astronomical Units. With the above system, it would take us over 2,100 years to reach Proxima Centauri, our nearest star.

Variations on this system include the use of multiple lasers, all focused on the one sail, and also multiple sails, one or more of which could be detached and used to reflect light back to the main sail and act as a brake. With present technology, the only photon space propulsion practical is by using sunlight. However, it has been suggested that it may be possible to make fine controls to the attitude of a nanosatellite, a satellite with a mass between one and ten kilograms, by using laser thrusters.

Propulsion by Laser Ablation

A laser can be used to heat the surface of a solid propellant, such as a sail, until it evaporates, sublimes or desorbs and generates a vapour or plasma jet. This process is known as ablation. As the gas leaves the surface, it provides the thrust, which is about a thousand times greater than that gained from photon propulsion. The laser can be onboard the spacecraft, or it can be remote, either in space or on the Earth. This concept could use lasers that are currently produced, which have an optical power of between a few watts to several kilowatts, rather than the several gigawatts that a photon propulsion system would need. The laser would probably be pulsed rather than continuous, as this would allow time for the spacecraft to move away from the exhaust so that a clear optical path between the laser and the spacecraft is maintained. If the sail were close enough to the Sun, the heat radiated from it could be enough to cause ablation and accelerate a sail.

A laser plasma thruster, which uses electrically charged lasers to produce a plasma from a solid fuel, has been developed. This type of system is called a chemically augmented electric propulsion system. The amount of thrust is tiny but it can be controlled and varied very accurately, and so could be used to make critical minor adjustments to the attitude of a spacecraft.

A proposal to use laser ablation space propulsion to move space debris either into graveyard orbits or so they burn up in the Earth's atmosphere has also been suggested, although this is not yet practical, and it is expected to be very expensive.

Plasma produced by a laser could be further accelerated by an electric field, in the same way as a Pulsed Plasma Thruster (PPT) that is currently used in space. Tests show that this hybrid system has an efficiency similar to a conventional PPT, but it has a much higher specific impulse, which means that it produces more thrust per amount of propellant.

Propulsion by Laser Gas Detonation

Another form of laser propulsion is to use the laser beam to rapidly heat a gas or a liquid. An air-breathing, laser-propelled flyer that uses this technology, called a lightcraft, has been tested in the Earth's atmosphere. The shape of the lightcraft channels air flow into a ring at the bottom of the craft, and, using mirrors, the laser beam is focused onto this ring of air. The air heats up rapidly and ionizes into a plasma. This causes a blast wave to be created which pushes the lightcraft upward. The laser is pulsed so there are many blast waves as the spacecraft rises. As the lightcraft reaches about Mach 1, the lightcraft begins to start operating in a similar manner to a ramjet engine. The highest this type of laser-propelled flyer has so far reached is just over 71 metres. When the lightcraft reaches a height when air will no longer be an effective medium, the laser will continue to be used to expand an onboard propellant. This would probably be either liquid hydrogen or nitrogen and would continue to propel the spacecraft.

Within the next ten years, it is expected that some form of lightcraft will be propelled to a height of ten kilometres to prove the concept of laser propulsion. The major problem for this project is funding. Most of the technical problems have already been overcome as a lot of the equipment required is in use around the world, or could be manufactured relatively easily. Within the same time scale, it could also be possible to launch a lightcraft from the surface of Mars into a low Mars orbit, where it could be collected and returned to Earth as a Mars sample return. The major benefit of this system is that although the laser and associated equipment would have to be taken to and set up on the surface of Mars, it could be left there and used many times as only the lightcraft and payload needs to enter orbit. As the surface of Mars has a much weaker gravity than the surface of the Earth, and also as the density of the Martian atmosphere is low, a surface launch from Mars requires much less energy than a surface launch from the Earth. In about 15 years time it should be possible, if funding into research and development is sufficient, to routinely launch 10–20 kilograms satellites from the surface of the Earth into a low Earth orbit using this method.

Microwave Sailing

Lasers are not just limited to visible light, and so there are infrared lasers, ultraviolet lasers and X-ray lasers. However, if microwaves are emitted the result is known as a maser from "Microwave Amplification by Stimulated Emission of Radiation". Masers were actually invented before lasers. As the manipulation of microwaves is a little simpler than for laser beams, maser sailing may be an easier method of beam propulsion than laser sailing. Maser transmitters are also a lot more efficient than laser transmitters. However, the beam from a maser spreads out more rapidly than one from an optical laser due to their longer wavelength, and so it would be a lot less effective over long distances than a laser. This means that the sail would have to be accelerated to the coasting speed quickly, before it moved out of range.

The principle of microwave sailing has been successfully tested in a laboratory, and a space-based test was to be conducted on the solar sailing spacecraft *Cosmos 1*. For the test, microwave energy was to be beamed from JPL to the spacecraft, and the microwaves would have hit the sail and pushed the craft forward. Unfortunately the launch vehicle taking *Cosmos 1* into space failed and the spacecraft never entered orbit.

The late Robert L. Forward was the first to propose a maser-pushed spacecraft, which he called a *Starwisp*. He suggested that they could be used as a fly-by mission to other stars. The sail could be composed of a wire mesh. As the wavelength of the microwaves is larger than the gaps in the mesh, it would appear as a solid sheet of metal to the beam, which would push on the sail and be reflected back in the opposite direction. The mesh would cover an area of a square kilometre, but would have to be made of micron-thin material and so it would weigh less than 30 g. The acceleration on the sail would be very high and could allow the spacecraft to reach a coast velocity of around 20% of the speed of light.

The microwave beam transmitter would need to be in an orbit close to the Sun, so that it could generate the required power through solar cells. The beam would also need to be focussed through a lens, onto the *Starwisp*. If tiny circuitry could be embedded into the mesh to act as the controls, sensors and cameras, when the *Starwisp* passed near to another star it could use the sail as an antenna to beam the images and information back towards the Earth. At this point, the microwave transmitter in our solar system could be switched on again and could provide enough power for the circuitry on the sail to operate. As there would be no method of slowing down the *Starwisp*, it would fly past the nearby star relatively quickly. However, once the microwave beam transmitter and the focusing lens were in place in the solar system, the cost of new *Starwisps* would be small, particularly compared to other interstellar options, and so many probes could be sent out to glean as much information as possible, before other, more controllable, probes were sent.

Masers can also be used for propulsion by ablation and a combination of solar sailing and maser ablation could be used to get a spacecraft into an interplanetary orbit. A suggested mission involves the spacecraft being deployed into a low Earth orbit by a conventional rocket. After the sail has been deployed, a microwave beam heats the surface of the sail, which ablates and the escaping gas propels the craft. Each time the craft passes over the maser transmitter, it is shot at again, therefore making the elliptical orbit more elongated and apogee further away from the Earth. At a certain point, it will begin to fall towards the Sun edge on. When it approaches the Sun, it will be turned so that the sail faces the Sun, and under the intense sunlight, the second surface ablates, boosting the spacecraft's speed. As it passes the Earth, it could receive another maser boost. By this stage all of the ablative material has been removed and the sail is then pushed by solar photons out through the solar system and beyond. Further research is required to determine the overall credibility of this type of mission.

Microwave Power

Masers could alternatively be used to power conventional electric propulsion systems. The energy from a maser could be converted into electrical energy by special antennas on a spacecraft. These are called "rectennas" or rectifying antennas. This electrical energy could then power a propulsion system or electronic equipment onboard the spacecraft.

Pellet-pushed Probes

Instead of electromagnetic radiation, very small pellets could be fired at a spacecraft and impart their momentum to push it along. This would overcome the major problem of electromagnetic beams, which spread out with distance and become less powerful.

The pellets could be launched from the Earth, probably by an electromagnetic mass driver, described later, and then accelerated when they were in the solar system by a space-based accelerator that would be powered by nuclear or solar energy. The stream of pellets would be carefully aimed at launch, but could also be realigned during their flight. Devices on the route could make coarse adjustments, either using electromagnets or electrostatic materials, and the fine adjustments could be made from a distance by photon pressure from a laser. When the pellets reach the spacecraft, they could be reflected back in the opposite direction, just like a ball bouncing off a wall or the pellets could be vaporized into a plasma, which would then be reflected by a magnetic field. In the future, a pellet launcher, or maybe even a reflector, could be set up at the destination to provide two-way travel.

Magnetic Sails and Plasma Sails

In theory, it is possible to harness the momentum from the solar wind. The solar wind consists of charged particles. By passing these through a magnetic field, they will be deflected and their momentum could be transferred to a spacecraft. One method that has been proposed to utilize this is for the spacecraft to deploy a large loop of superconducting wire, or a coil of loops, called a solenoid, to generate the magnetic field. The wire only needs to be a few millimetres in diameter. Other loops could possibly be used for steering or to reduce the radiation hazards from the charged particles. There are still some technical problems to be solved before this type of propulsion system could be used, including further research on superconductors and the controls required to keep the ring at a low enough temperature for operation. It has also been suggested that instead of a loop of superconducting wire, a positively charged sphere, surrounded by a negatively charged rotating ring, is used, in a system called an electrostatic MagSail. However, further theoretical work is required to determine the amount of thrust that would be gained from such a system.

By injecting plasma into a solenoid, a magnetic bubble is formed, which could be used to form a magnetosphere around a spacecraft, similar to the one around the Earth. The solar wind pushes on the Earth's magnetosphere constantly and can be seen as the aurora. If enough solar wind hit a large enough bubble around a spacecraft, it could be used to propel the spacecraft along. To make the bubble bigger, all that is required is to inject more plasma near to the solenoid, just like blowing up a balloon. This type of propulsion system is called Mini-Magnetospheric Plasma Propulsion (M2P2). The size of the bubble would contract or expand, depending on how much solar wind reached it. The solar wind reduces with distance from the Sun, but as it reduces, the bubble will get larger, and so the spacecraft would receive a constant thrust, no matter where it was in the solar system. The major advantage of this system is that the plasma generator and the magnet only requires a small amount of energy, which could be provided by solar cells. The formation of a plasma bubble has been tested in a vacuum chamber on the Earth, but further research and modelling of the system is required before a system could be tested in space. If the plasma bubble contained dust particles, the bubble could maybe also harness the solar radiation or photon energy, and therefore accelerate the spacecraft faster.

Tethers

The stabilization and attitude control of satellites using a long cable, a boom or a tether and also the Earth's gravity is called gravity gradient stabilization and is described in Chapter 2 – "Rockets and Spacecraft". These designs use the Earth's gravity, or the gravity of the body the satellite is orbiting, to keep the satellite's long axis pointing towards the centre of the Earth, without the use of electronics or fuel. Gravity gradient stabilization is also used to link constellations of satellites together, so that they remain in a stable configuration. However, tethers can also be used a means of propulsion.

Electrodynamic Tether Propulsion

A tether incorporating an insulated conductive material, such as aluminium, throughout its length is called a conductive tether. As it orbits the Earth vertically a current can be driven through the tether. The ionized gases in space, such as from the solar wind, are used to complete the circuit. The electricity can be generated from a renewable source such as solar cells and therefore can continue indefinitely. As the tether crosses the lines of the Earth's magnetic field, a Lorentz force is produced, which pushes the tether, and any attached spacecraft, into a higher orbit. The Lorentz force is a thrust produced as the result of the interaction of an electric current and a magnetic field, and is described in detail in Chapter 5 – "Propulsion Systems". If the direction of the current is reversed, the force will be in the opposite direction and the orbit will be lowered. Instead of acting like a motor, this type

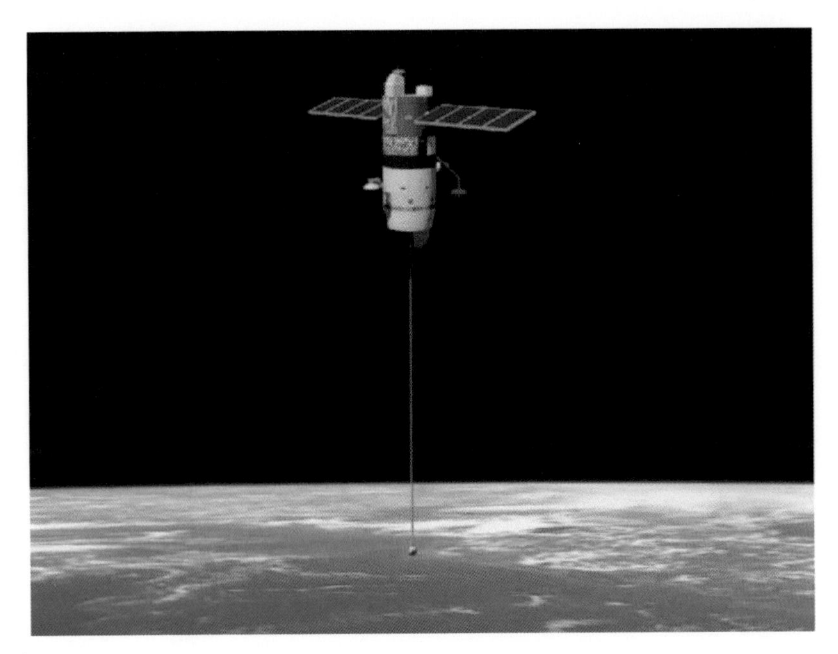

FIGURE 11.2 Artist's Impression of an Electrodynamic Tether Propulsion System.
Image courtesy NASA Marshall Space Flight Centre (NASA-MSFC)

of tether can also act as a generator to generate electricity for use within the spacecraft, without the use of solar cells. Figure 11.2 shows an electrodynamic tether propulsion system.

Rotating Tether

A tether could also be used to change the speed of an object by transferring the angular momentum of one object to another object. Angular momentum is a measure of the amount of motion an object has travelling around a point. There are various ways that momentum can be transferred. The simplest is by one moving object hitting a stationary one. This can be seen in the children's game of conkers. If a stationary conker, or horse chestnut, is dangled from a string and is hit by another conker, the stationary conker swings away and the moving one is slowed down.

Rapidly spinning asteroids or moons could be used, with the help of tethers, to eject materials and move them across space.

Therefore material mined from the asteroid or moon could be moved to other places, such as into an Earth orbit. One end of the tether would be connected to the equator of the moon or asteroid. The material would then either be attached directly to or propelled along the cable to the free end. As the cable spins, the material speeds up and if the cable were long enough, when the material is released from the end, it would fly off into space, as the gravity of the body would be insufficient to keep it in orbit. This would only work in a vacuum, as the friction from any atmosphere would cause the cable to burn and break.

Instead of using the spin of the asteroid or moon, a rotating hub attached to the surface at either the north or south poles could be used instead. A cable, attached at one end to the hub, would lift up from the surface as the hub rotated. A payload attached to the free end of the cable could then be accelerated and released, just like an athlete's hammer throw. Although this system would use a lot of energy to rotate the hub, the rotating tether system has the advantage of simplicity and reusability. This type of system has not yet been used in space, but organizations such as NASA, Tennessee Technological University, Lockheed Martin Astronautics and Tethers Unlimited are developing tether systems for use, hopefully, in the near future.

Rotating Sling and Momentum-Exchange Electrodynamic Reboost (MXER) Tether

In a rotating sling system, a very long thin tether rotates about a large mass connected at its centre. The tether would probably need to be over 100 kilometres long. The tether and mass need to be in an elliptical orbit about a body, for example, around the Earth, Moon or Mars. The system is timed so that when the tether is nearest to the body the tether must be aligned vertically below the mass and swinging downwards. A payload or spacecraft will rendezvous with the tip at this stage and link to it. As the tether continues to rotate, it takes the payload with it, and when it has completed half a rotation, it releases it. The payload is now in a higher energy orbit, that is, it reaches a higher altitude from the body it is orbiting.

The tether, however, now has a lower energy orbit and so it does not reach such a high altitude, as it has transferred some of its momentum to the payload. To enable the tether to boost another payload, it must be recharged so that it is back in a higher energy orbit. This can be done without propellants by using electrodynamic tether propulsion. This type of system is known as a Momentum-Exchange Electrodynamic Reboost (MXER) tether propulsion system. Due to the limitations of current materials and also the challenges involved, including the linking or capture mechanisms, these types of system have not yet been used in space.

Tether Payload Deployment

Another form of momentum exchange is tether payload deployment. This is where one spacecraft, for example, the Space Shuttle, deploys a payload on a tether while in orbit around the Earth. As the payload and spacecraft are connected, they orbit at the same speed, even though they are at different altitudes. If the payload is deployed down towards the Earth, the payload travels a little slower than it would normally at that altitude and the spacecraft a little faster. If the tether is then cut, the spacecraft will be boosted into a higher energy elliptical orbit, and the payload would drop to a lower one. The point where the tether is cut will be the perigee, or point closest to the Earth, on the new orbit for the spacecraft and the apogee, or point furthest from the Earth for the payload's new orbit. A similar method could be used to boost a payload into a higher orbit. By careful deployment, the change in the orbit may benefit both of the orbiting objects, in an application called momentum scavenging. This can only be achieved if there is an excess of momentum. An application of momentum scavenging could be a Space Station – Space Shuttle deboost operation. When the Shuttle is ready to leave the Space Station, it could undergo a tethered deployment. When the tether is released, the Space Station would be boosted to a higher orbit, whereas the Shuttle would go to a lower orbit in preparation for its return to Earth. This system is still only at the research stage and has not yet been tried.

Space Elevators

The idea of using a ladder or tower to reach the sky or heavens has been around for a long time. Some of the earliest known manuscripts include ideas such as the Tower of Babel and Jacob's Ladder, although these were never practical. The idea has probably been reinvented many times, however, it was not until 1975 when Jerome Pearson published a paper on the subject, that it was brought to the attention of the space flight community. Sir Arthur C. Clarke wrote about one in his 1979 novel *The Fountains of Paradise* and NASA started research work on space elevators in 1983.

There are numerous designs of space elevators, also referred to as beanstalks, spacebridges, space lifts and space ladders. The basic principle of a space elevator requires a physical connection between the Earth and space. One idea is for a cable, ribbon or tether to stretch between the Earth's equator and a station in orbit. An elevator or climber would ascend the ribbon, taking payloads or human cargo with it. The climber would not use rocket propulsion, but probably use electricity to power an electro-mechanical or electromagnetic drive system. It is thought that this type of system would bring the cost of transporting items into space to similar levels as currently transporting items by air. The centre of mass of the system would be in a geostationary orbit and so it would appear to hover over the same part of the equator all the time. If the orbit were any higher or lower, the centre of mass of the system would travel at a different speed than the Earth rotates and therefore the ribbon would stretch and break. If the base station was located on a floating moveable platform in the sea, the system could possibly be moved to avoid meteoroids, space debris and other satellites.

A second idea is for a tower to be built from the Earth that is so tall it reaches into space. Currently, the tallest man-made structure on land is the KVLY-TV mast in North Dakota, USA, which stands at 628.8 metres. The Burj Dubai, in Dubai, United Arab Emirates, which is currently being built, is expected to be about 800 metres high. With the technology and materials used today, it is already possible to build structures many kilometres in height, although nobody has yet had a compelling reason to do so. A combination of a ribbon and a tower will provide a good solution to the structural strength requirements of an elevator system. It is not yet possible to build such a system with current technology, mainly because suitable materials for the cable are not available. However, even if they were, the design would be so large and costly that it would not be viable. New materials, such as carbon nanotube ropes, are being developed which may make the development of a space elevator feasible within the next 50 years. Other technological advances that may be required for a space elevator include the development of electromagnetic propulsion, which would be the ideal solution for climbing the ribbon. As this type of propulsion does not require contact with the ribbon, the ribbon would not wear through friction and would therefore last longer.

Space Fountain

This is an idea for a space elevator that does not rely on a geosynchronous orbit or the strength of the material it is built from. Instead, it is a very tall tower that is held up by kinetic energy. Massive pellets are fired upward from the bottom of the tower and when they reach the top, they are reflected or redirected back down. The force of the redirection holds the top of the tower up, in a similar way that a ball can be held aloft on a jet of water. The payload can either be somehow attached to the stream of pellets, or could be lifted up the side of the tower. A space fountain has the advantage over the elevator in that it can be located anywhere on the surface of the planet and it can be raised to any height and does not require very strong materials. However, it does require a constant power supply to remain upright, and once it had reached the top, the payload would need to be propelled to orbital velocity by another method, for example, by a propulsion system on the spacecraft. Instead of using pellets, it has also been suggested that a closed-loop cable is used to travel up and down the tower and hold it upright. A space fountain is theoretically feasible, but it would require a lot of engineering development and a large financial commitment to build and maintain. It would also be expensive to run, in terms of energy. Therefore other, less demanding, projects will probably be investigated and funded before this type of system receives much more attention.

Sky Hooks

An intermediate step towards the geosynchronous elevator is to place a station in a low Earth orbit with a tether dangled down, pointing to the centre of the Earth, and a second tether pointing away from the centre of the Earth. The end of the lower cable would be about 100 kilometres above the Earth. It is the centre of gravity of the cable system, called the midpoint station that determines the orbital velocity of the cables. The bottom of the cable is therefore travelling slower than orbital velocity for its height, and the top of the cable is travelling faster. This means that the payload that is transferred to the bottom of the cable does not have to reach orbital velocity from launch. A space plane could be used to transfer cargo from the surface of the Earth to the bottom of the cable. The space plane would probably have to be more powerful than those currently planned for sub-orbital space tourism. The cargo could then be raised up the tether using some transport mechanism such as a crawler, either to the midpoint station or to a parking orbit higher up the cable for transfer to a journey to the Moon or beyond. As the top of the cable is travelling faster than orbital velocity for that height, a payload released from here will receive a free velocity boost. As the lower end would not be fixed to the Earth, it can be placed on an inclined orbit, aligned with the ecliptic. This would be beneficial for travel out to the Moon or to other planets, as the orbital inclination would then not have to be changed. The sky hook would have to be reboosted to a higher orbit regularly, as it will be slowed down by the payload and also by the very thin atmosphere that it will pass through.

Nuclear

There are potentially three different types of nuclear energy source available, fusion reactors, fission reactors and radioisotope decay. Nuclear fusion happens when two or more atoms combine to form one heavier atom. This is usually accompanied by a huge amount of energy, mainly in the form of heat. The Sun produces heat by nuclear fusion, where hydrogen atoms combine to form helium atoms and energy. Different concepts have been considered for nuclear fusion propelled rockets, but none have been tested, as there is no known practical method of harnessing any useful energy from the fusion reaction.

The heat from a nuclear fission reaction could be used to heat a propellant, such as liquid hydrogen, so that it expands and is expelled through a nozzle to provide thrust. Fission reactor rockets were designed and tested on the Earth in the 1960s in the USA. However, due to concerns over safety and the practicalities of the endurance of the materials at such high temperatures, along with no realistic goal of a manned Mars mission, the program was shut down.

The heat energy produced from the radioactive decay of a nuclear source such as Plutonium 238 could also be harnessed and used for direct heating of a propulsion fluid, which would then be expelled through a nozzle. As the power produced from this would be quite small, an isotope decay engine could not be used as a high thrust engine but could instead be used for low thrust space manoeuvres. However, this type of engine has not yet been developed or flown, probably because the amount of power generated by such an engine is tiny, especially when compared with a nuclear fission reaction, which itself is small compared to the energy of a nuclear fusion reaction.

Radioactive decay can also be harnessed when used on a sail. A radioactive chemical could be painted onto one surface of a sail. Some of the radioactive particles emitted by the chemical would be absorbed in one direction by the sail, but would be free to leave the sail in the other. They would therefore provide a thrust and push the sail along. This type of sail could only work in space, as the resistance from the atmosphere would prevent the sail from moving. By moving the radioactive source from the surface of the sail to the centre point of a hemispherical shaped sail, particles that were emitted towards the sail would heat the sail and, if the back of the sail were insulated, it could produce a further thrust by photons leaving the sail as infrared radiation. Another design is to use an electrostatic reflector, which directs the charged radioactive particles into one direction. A combination of the two could also be used.

Magnetic Satellite Launch System or Magnetic Mass Drivers

Earth based transport systems that used a magnetically levitating (maglev) vehicle were patented over 100 years ago. However, development of maglev vehicles has developed slowly, and it was not until 1984 that the first commercial maglev was opened, linking the airport in Birmingham, England to a rail station 600 metres away. However, the technology is not limited to train travel. Studies have shown that the use of magnets to accelerate a vehicle along a straight track can produce the high speeds required for a launch into orbit. This straight-line method, however, means the vehicle must gather speed in one quick burst, which uses a huge amount of energy for a very short time. Obtaining enough energy is difficult. A further idea is to use an enormous ring of superconducting magnets and the spacecraft could be gradually accelerated as it goes round and round, over a period of hours. This evens out the energy requirement to more accessible levels. The spacecraft would have to be on some form of carrier or sled, which would be accelerated by the superconducting magnets. When it reached the desired speed, it would separate from the sled and travel up a ramp and be launched into space. A rocket at the back of the spacecraft could be used to adjust the trajectory and place it in a proper orbit. Figure 11.3 shows a potential magnetic launch assist system.

The spacecraft would probably launch at a very high *g* force, and so it would not be suitable for sensitive equipment or manned vehicles. It would also have to be encased in an aerodynamic shroud, to protect it from the intense heat during launch caused by friction through the atmosphere. The system could be used to deliver supplies to support human spaceflight, such as food and water, which are not sensitive to such high accelerations. This type of system could also be used to propel objects from the surface of other bodies, such as the Moon, where it would not have to overcome atmospheric drag and the gravity is less.

This system is technically feasible, although much of the technology needs further development. If enough items are launched every year, it could be cheaper than launching items using conventional rockets. However, getting the spacecraft safely through the

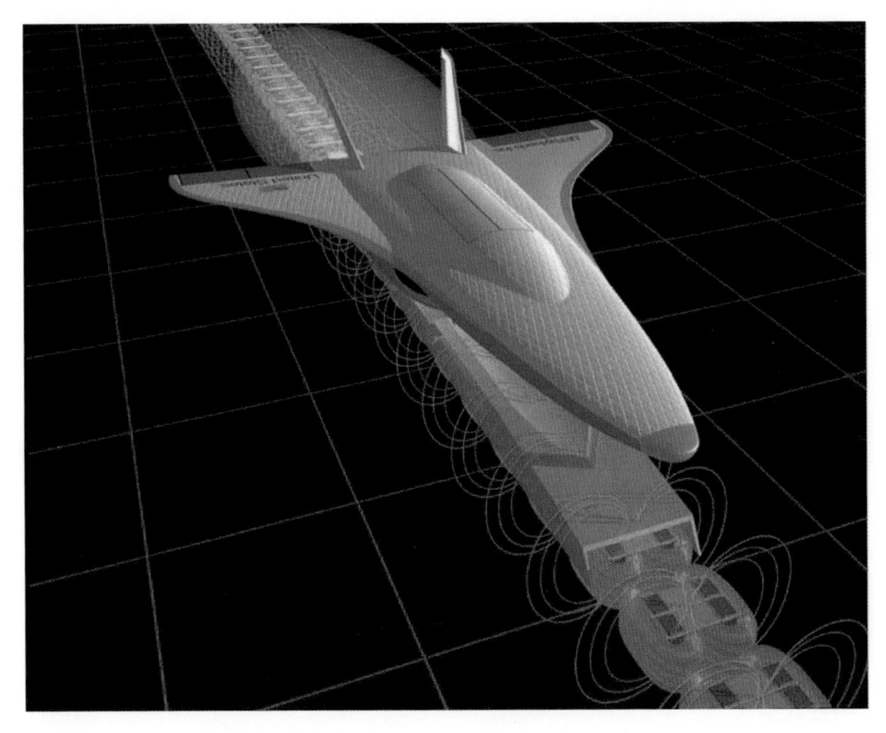

FIGURE 11.3 Artist's Impression of a Magnetic Launch Assist System. Image courtesy NASA Marshall Space Flight Centre (NASA-MSFC).

atmosphere could pose other challenges. For example, wind and the atmosphere could cause the spacecraft to veer off course.

The Distant Future

There are many ideas for travelling through space that, with current scientific knowledge, are impossible. These include anti-gravity devices and methods used in the fictional Star Trek programs, such as teleportation, faster-than-light travel by warp drives and bending space though wormholes. If we could break our scientific rules and develop these types of technologies, we would be able to travel to the stars faster, but even without them, the exploration of the Universe is still possible. All that is required to get from here to there are the vision, determination and finances to make it possible. As it says in *The Global Exploration Strategy: The Framework for Coordination*:

> The human migration into space is still in its infancy. For the most part, we have remained just a few kilometres above the Earth's surface – not much more than camping out in the backyard. It is time to take the next step.

Appendix A: Orbital Elements

Many of the items in this Appendix have been explained elsewhere in the text, but, for clarity, they have been drawn together into this one section.

The size and shape of a satellite's orbit is determined by its speed and mass, and also by the mass of the object it is orbiting. The size, shape and orientation of an orbit can be described by six orbital elements or orbital parameters, known as Keplerian elements or element sets. Different sets of elements can be used for different purposes. The terms used sound complicated, but taken individually each is relatively simple. These explanations have been taken with reference to an Earth-orbiting satellite, however, they can easily be adapted for satellites orbiting other celestial bodies. If all of the elements are known, the exact location of a satellite can be calculated.

The Keplerian elements are:

Inclination (i)
Longitude of the ascending node (Ω)
Argument of periapsis (ω)
Eccentricity (e)
Semi-major axis (a)
Anomaly at epoch (v)
Time of periapsis or perigee passage (T) may be used instead of the Anomaly at epoch (v)

Inclination (i)

A satellite will always travel around the centre of the body it is orbiting and therefore it will always cross the equator twice in every orbit, unless, of course, it orbits directly above the equator. It will cut it as it travels from the southern hemisphere to the northern and again when it crosses from the northern to the southern hemisphere. The angle it makes when it crosses the equator from the southern hemisphere to the northern hemisphere is called the inclination, and this defines the orientation of the orbit with respect to the equator. Therefore, an orbit that is directly above the equator has an inclination of 0° and one that goes directly over the north and south poles has an inclination of 90°. In Figure 1.2, the orbit shown in red has an inclination of 28°, which is the inclination of most of the USA's scientific satellites. The yellow orbit shows the path of the International Space Station, which has an inclination of about 52°.

Longitude of the Ascending Node (Ω)

The ascending node is where the orbit of a satellite crosses the Earth's equator when the satellite is travelling from south to north, as shown in Figure A.1. The longitude of the ascending node is defined with reference to the terms celestial sphere, the ecliptic and the First Point of Aries, which therefore need to be defined first:

The celestial sphere is the imaginary sphere encompassing the Earth, as can be seen in Figure 6.5. All objects in the sky can be thought

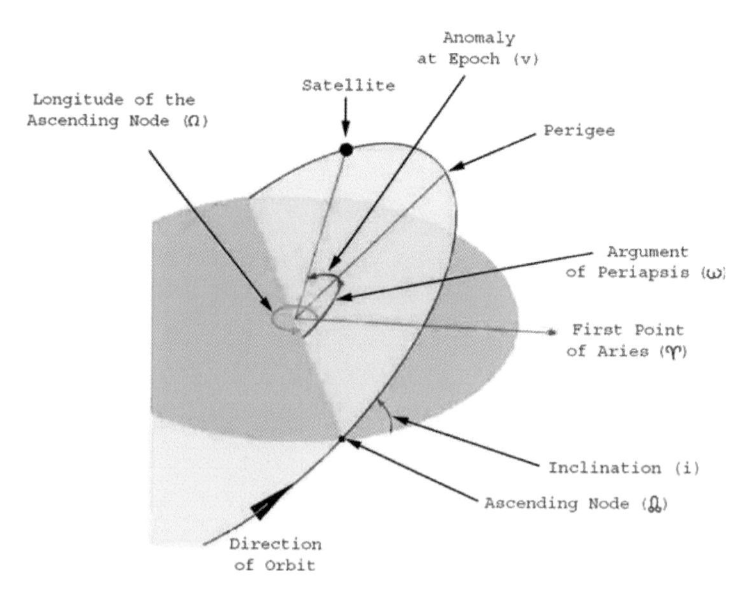

FIGURE App A.1 Orbital Elements

of as being on the sphere. The celestial equator and the celestial poles are projected from their corresponding earthly equivalents. Angles measured north or south from the line of the celestial equator are called the declination and are measured in degrees. This is similar to lines of latitude. The longitude equivalent is called the Right Ascension or RA. This is measured in hours eastwards from the First point of Aries, which is given the symbol ♈.

The Sun does not always appear directly above the equator but instead follows a path called the ecliptic, which takes it above and below it. When its path crosses the equator, it is called an equinox. For the northern hemisphere, the Vernal Equinox is usually about March 20. This is when the length of daylight and darkness is exactly equal, and the Sun is directly above the equator. The direction of the Sun from the Earth on this day is known as the First point of Aries.

The longitude of the ascending node is the angle from the First Point of Aries to the ascending node, taken anticlockwise if looking down from the north, in the plane of the Earth's equator.

Argument of Periapsis or Perigee (ω)

This defines where the perigee of the orbit is, in respect to the Earth's surface. This is the angle formed, in the plane of the satellite's orbit, from where the satellite crosses the Earth's equator at the ascending node, to when it reaches perigee. If the angle is less than 180°, perigee is in the northern hemisphere, larger than 180°, it is in the southern hemisphere.

Eccentricity (e)

This is the shape of the orbit. Most orbits are not circular but look liked squashed circles, called ellipses. How flat the ellipse looks is called its eccentricity and it is given a value from zero to one. An ellipse with zero eccentricity is a perfect circle and a very flat ellipse has an eccentricity nearing one, as can be seen in Figure 1.3.

Semi-major Axis (a)

This defines the size of the orbit. The long axis of an ellipse is called the major axis, the shorter one, the minor axis. In a circle, both the major and minor axes are the same, and are called the diameter. Half of the major axis is called the semi-major axis. This is the equivalent of the radius in a circle. The length of the semi-major axis is used to describe the size of the ellipse. The properties of an ellipse are shown in Figure 4.4.

Anomaly at Epoch (v)

Anomaly in this sense is an angle. The anomaly at epoch is the angle between where the satellite is now, compared to where it was at perigee. If the satellite is at perigee, the anomaly is zero. At apogee, it is 180°, and back to perigee is 360°. As most orbits are not circular, the measurement of this angle is not easy and so the angle is given in terms of a fraction of an orbit, with one orbit being 360°.

Time of Periapsis or Perigee Passage (T)

This is simply the time when the satellite was at perigee. This usually comprises the date and time, given in Universal Time or UT. For most purposes, this can be taken to be the same as Greenwich Mean Time (GMT).

Appendix B: Coordinate Systems

Geocentric Coordinate Systems

Usually, the centre of the Earth is used as the origin for the coordinate system for spacecraft in orbit around the Earth. This is called a geocentric coordinate system and is shown in Figure B.1. A geocentric equatorial coordinate system is a Cartesian system that has the fundamental plane as the plane of the equator and the x direction towards the Vernal Equinox or First Point of Aries. The z direction is through the North Pole and the y direction makes up the right-handed set. Except for precession, the wobbling motion of the Earth's axis of rotation, this system is fixed to the stars and the Earth rotates within it.

Heliocentric Coordinate Systems

For satellites travelling to other places within our solar system, a coordinate system with the origin at the centre of the Sun is used, such as the heliocentric ecliptic coordinate system. This can be seen in Figure B.2. The fundamental plane is the plane of the ecliptic, with the x direction being towards the Vernal Equinox or First Point of Aries at a particular epoch. The z direction is the ecliptical North Pole, and the y direction makes the right-handed set.

Perifocal Coordinate System

A perifocal coordinate system is often used to describe the motion of a satellite. The fundamental plane is the plane of the satellite's orbit, and the x-axis points towards the part of the orbit nearest to the body the satellite is orbiting, known as the periapsis. The y-axis is on the orbital plane, and rotated 90° from the x-axis in the direction of the satellite's motion. The z-axis completes the right-handed set.

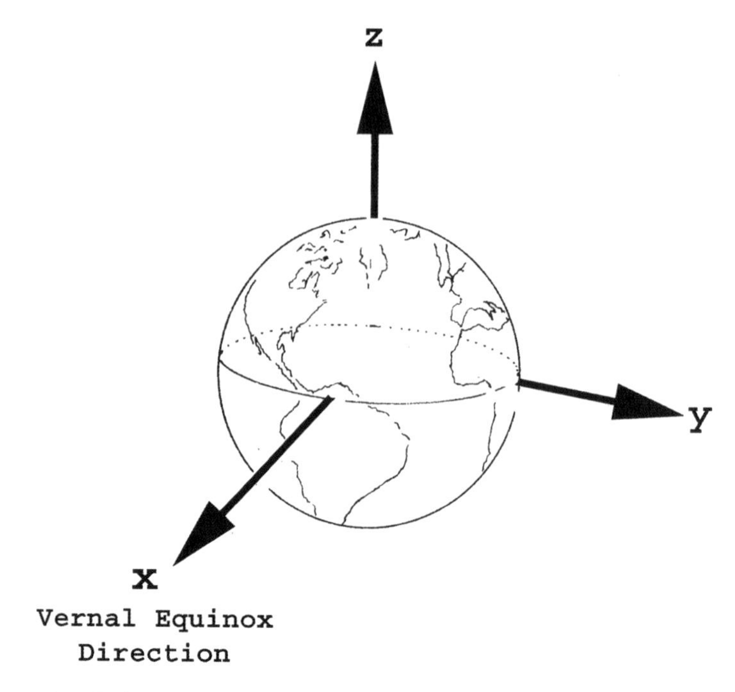

FIGURE App B.1 Geocentric Equatorial Coordinate System

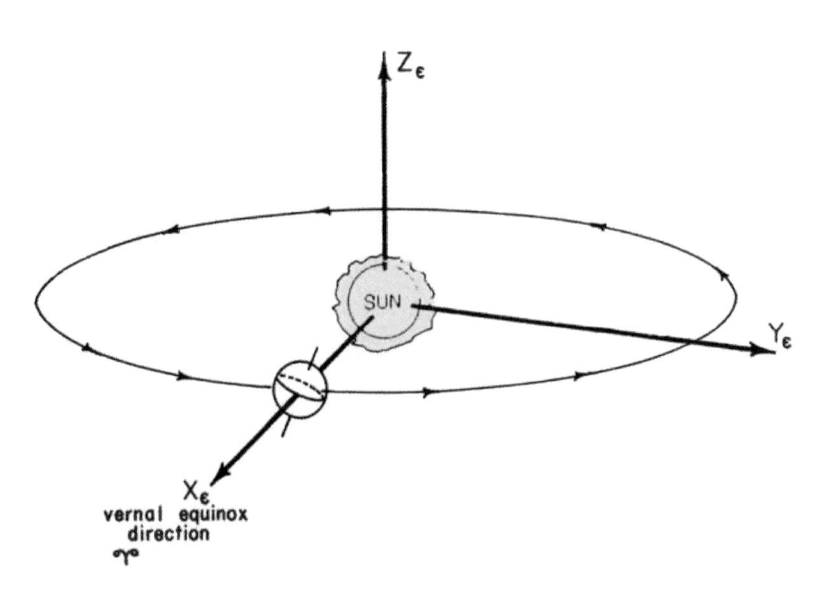

FIGURE App B.2 Heliocentric Ecliptic Coordinate System

Appendix C: Web Site Addresses

Name	Web site	Comment
The Radio Amateur Satellite Corporation (AMSAT)	www.amsat.org	An educational organization with the goal of fostering amateur radio's participation in space research and communication.
Amateur Radio on the International Space Station (ARISS)	www.ariss-eu.org	Common working group of the national amateur radio and amateur satellite radio societies involved in amateur radio operations on board the International Space Station.
Belgian Working Group for Satellites	www.vvs.be	Group that coordinate and collate amateur's data about the flash period of satellites. The web site is in Dutch.
Celestrak	www.celestrak.com	Dr. T.S. Kelso's web site, which provides a lot of satellite tracking and other information.
ESA	www.esa.int	Web site of the European Space Agency.
HearSat	www.hearsat.org	Satellite radio signal monitoring web site.
Heavens Above	www.heavens-above.com	Provides information needed to observe satellites as other spaceflight and astronomical information.

(continued)

Name	Web site	Comment
International Meteor Organization (IMO)	www.imo.net	Ensures international cooperation of meteor amateur work. Collects meteor observations by several methods from around the world.
ISS Fan Club	www.issfanclub.com	Provides realtime information about the International Space Station. Most of the content is related to amateur radio. It is not affiliated to any official body or organization.
Mike McCants	www.io.com/~mmccants/	Mike McCants' satellite tracking web pages.
NASA	www.nasa.gov	NASA's web site.
NASA ISS Predictions	spaceflight1.nasa.gov/realdata/sightings/	Sighting opportunities of the ISS, provided by NASA.
Space Track	www.space-track.org	Source for space surveillance data. Registration is required to access the site.
Spacewarn Bulletin	nssdc.gsfc.nasa.gov/spacewarn/	Monthly publication for satellite information.
Sven Grahn's Sounds from Space	www.svengrahn.pp.se/sounds/sounds.htm	Radio signals from space.
The Organization for Weather Satellite and Earth Observation Enthusiasts	www.geo-web.org.uk	Provides help for amateur reception of weather and Earth imaging satellites which are already in orbit or planned for launch in the near future.
Visual Satellite Observer's home page	www.satobs.org	Provides information on all facets of visual satellite observation.
Zarya	www.zarya.info	Information on space programmes and space research undertaken by the Soviet Union and Russia.

Appendix D: Practical Information for Observing Satellites

The information given by most prediction sites includes the name of the satellite, the date and time when the satellite will appear, along with the altitude and azimuth. The satellite name is usually given in one of three ways, the common name, the International Identification Number or the Space Catalogue number. The owner of the satellite usually assigns the common name, such as the ISS, *Iridium 28*, the *Hubble Space Telescope* or *Chandra*. Some names can cause confusion as many satellites have multiple payloads aboard. Unique identifiers are therefore often used to refer to a specific satellite. These are the International Identification Number and the Space Catalogue Number.

In satellite predictions for a specified location, the altitude is the height above the horizon. It is usually given as an angle, with 0° being on the horizon and 90° being directly above the observer. A fist held at arms length is approximately 10° wide. The azimuth is the direction east or west, and can be given as angle, with 0° being due north and 90° being due east. It can also be given as a compass direction. Real north and not magnetic north is used. The location of Polaris, the Pole star is therefore useful to find if observing from the northern hemisphere, as it is less than 1° off true north. Unfortunately there is not a similar star for the southern hemisphere, but by location of the constellation the Southern Cross, the approximate position of real south can be determined. Alternatively, the difference between magnetic north and true north can be calculated. The difference varies around the globe, and there are web sites that provide the calculation if the location's coordinates are given.

Right Ascension or RA and declination or Dec are similar to the longitude and latitude on Earth, but are measured from the celestial equator. Declination is measured in degrees, and refers to how far above or below the celestial equator an object is. Right Ascension is measured in hours of time, and is measured clockwise around the celestial sphere, with zero hours being in the direction of the Vernal Equinox or the First Point of Aries.

Two-Line Element (TLE) Sets

More detailed information for the prediction of satellites is often in the form of Keplerian element sets, as described in Appendix A – "Orbital Elements". The element sets are also known as Two-Line Element (TLE) sets, Elsets or ephemeredes. This information gives all of the data regarding the orbit of the satellite and, from this, it is possible to determine when a satellite will pass overhead, the direction of travel and how long it will remain illuminated. There are many pieces of software available on the internet for translating the orbital parameters into data that amateurs can easily use.

The TLEs are usually given in either the NASA format or the AMSAT format. The NASA format is the type provided by NASA for all its satellite element data, an example of which is given in Table D.1. It provides extensive information about the orbit, some of which is not usually required by an amateur. Despite being called two-line elements, the data is actually provided in three lines. The first row gives the satellite name, and is referred to as line zero. Lines one and two are labelled as such. There are 69 columns, and the information in each is for a specific piece of data, so the spaces are important.

In row 1, column 1 contains the row number and columns 3–7 are used for the Space Catalogue number. Column 8 is a classification.

Table App D.1 NASA Format Files

```
*         1         2         3         4         5         6
123456789012345678901234567890123456789012345678901234567890123456789
ISS (ZARYA)
1 25544U 98067A   07200.28502481  .00016615  00000-0  94107-4 0  6136
2 25544 051.6358 297.9340 0008913 080.2435 014.8514 15.79196610495845
```

*The top two lines are not normally given, but are included here for clarity.

Most TLEs available for public use will contain a "U" for unclassified in this column. Columns 10–17 are the International Identification Number, 19–32 contain the date and time, or epoch that these elements were calculated for. Columns 34–43, 45–52 and 54–61 are used to calculate the influence of drag on the orbit. Column 63 is the ephemeris type, which can be used to distinguish which method was used to generate the TLEs. Different methods are used depending on the accuracy required. For example, for precise analyses, such as re-entry predictions, a "Special Perturbation" theory is used, which uses high precision gravity models, atmospheric density drag models, three-dimensional body perturbation models and other precise numerical methods. For general ephemeredes, including most of those that are made available to the public, a "General Perturbation" theory is used. This field is usually left blank or set to zero in the publicly available TLEs. Columns 65–68 contain the element number or how many TLEs have been generated for this object, although a few numbers are missed to avoid ambiguities between the different agencies that produce the TLEs. After 999 this figure reverts back to zero. The last number is a check sum, and is used to help ensure the data is correct.

In row 2, again column 1 contains the row number, and columns 3–7 contain the Space Catalogue number. Columns 9–6 are used for the inclination in degrees, 1–25 for the Right Ascension of the ascending node in degrees, 27–33 shows the eccentricity, and 35–42 the Argument of periapsis in degrees. Columns 44–51 give the Mean Anomaly in degrees, 53–63 the Mean Motion, which is an average of how many orbits the object makes per day. Columns 64–68 show how many revolutions the orbit has made since the launch to the epoch date. The last digit is again a check sum for the row.

The AMSAT format covers various similar formats that are produced by different people or groups, an example is given in Table D.2. Most tracking programs try to read all of the different types. This format is very user-friendly, and is much easier to read and understand than the NASA format. Each orbital element appears on a separate line, the order of which is not significant. Every new element set should begin with a line containing the word "satellite" and a blank line is usually used to end the set.

Table App D.2 AMSAT Format Files

Satellite:	AO-10
Catalog number:	14129
Epoch time:	07192.34214358
Element set:	0399
Inclination:	026.3670°
RA of node:	267.5095°
Eccentricity:	0.6065144
Arg of perigee:	037.9383°
Mean anomaly:	352.1297°
Mean motion:	02.05867411 rev/day
Decay rate:	3.5e-07 rev/day^2
Epoch rev:	18106
Checksum:	303

Positional Observations

Positional observers measure the position of a specific satellite at a specific time. The easiest way to determine the position of a satellite is when it passes between two relatively close stars, at most about half a degree apart. The stars should not be too far apart from each other, as the satellite's position needs to be estimated to within a few arcminutes. The most accurate measurements are possible when the line connecting the two stars is at right angles to the path of the satellite. The distance between the two stars should be subdivided, with the total distance being "1", and half way being "0.5" so that if the satellite does not pass directly between the two, it can be easily estimated as to how close it was to one or the other. The position of a satellite can be determined if it passes in front of a star. However, if the star and satellite are not the same brightness, it can be difficult to get an accurate measurement because the glare caused by the brighter object obscures the dimmer object.

To get the time and position measurement, a stopwatch is started when the satellite intersects the line joining the two stars. The stopwatch is stopped when a time signal, for example, from a radio, registers a full minute. By subtracting the time on the stopwatch from the time signal, the exact time that the satellite crossed the imaginary line is gained. The time should be recorded with the highest possible accuracy. As human reaction time is about 0.1–0.2 seconds, the timing accuracy should also aim to be between 0.1 and 0.2 seconds precision.

How quickly the satellite crosses the sky, or its angular velocity, also determines the accuracy of the observation. A satellite in an orbit below about 500 kilometres moves across the observer's sky with an angular velocity of 0.5–1° per second. Therefore, in 0.1 seconds, which is the maximum accuracy a person can time, the satellite will have moved between 0.05° and 0.1° or between three and six arcminutes. This is the positional accuracy. As the angular velocity for higher objects is smaller, the positional accuracy for these satellites will be better than for lower orbiting objects.

The next stage is to plot the satellite on a star chart. First, the reference stars must be found. By subdividing the distance between the two stars using a ruler, the satellite's position can easily be plotted at the estimated distance from one or the other. The coordinates for the satellite can then be determined from the plotted position. It is also possible to calculate the satellite's position mathematically from the coordinates of the reference stars using a process called linear interpolation. Computer programs have been written that enable the precise coordinates of the satellite to be calculated by distinguishing the reference stars by using a simulated binocular or telescope field of view and entering the observed geometric and positional data. The programs then automatically produce a formatted observation report.

The results of observations are usually analysed by computer. Therefore a systematic and accurate reporting format is required, ideally one that can be read by both humans and machines. There are three formats in general use. For each format, the following information is usually required: Satellite code, using either the Space Catalogue number or the International Identification Number; the time in UT; the position of satellite in Right Ascension and declination; an estimate of the accuracy in time and position; the observer's exact position in latitude, longitude and also height above sea level; and the type, make and size of the equipment used. The magnitude or brightness of the satellite can also be included, although this is optional.

Flash Period Observations

If a satellite is rotating, it will appear to flash as the Sun reflects off different parts. By measuring the time between flashes, the speed of the rotation can be calculated.

Types of Reflection

When light reflects off a smooth surface, such as a solar panel or polished antenna, most of the light reflects in the same direction, and the reflections can reach negative magnitudes. This is called secular reflection. However, reflections off a microscopically rough surface, such as unpolished metal, spread out in many different directions, and this is called diffuse reflection.

Synodic Effect

The brightness of flashes from many satellites changes as they track across the sky. These flashes are caused either by the satellite's rotation, or by the synodic effect. The synodic effect is when the Sun is reflected from different parts of the satellite as the angle between the Sun, satellite and observer changes as the satellite crosses the sky. This synodic effect can easily cause confusion and can affect the accuracy of any rotation determination, and observers must be aware of it. The synodic effect may make the difference between the flash and rotational periods larger than the accuracy of the observations. Generally, this effect will be larger for lower objects. This is because they move faster across the sky and the geometry between the satellite, Sun and observer changes more drastically during one rotation period. The satellites with a longer rotation period also usually have a larger synodic effect, again because the geometry changes relatively quickly in relation to the rotation. Satellites with a rotation period below about ten seconds are usually not very much influenced by the effect.

Measurement of Flash Period

Flash periods are defined as the time interval between two flashes. Measuring the time interval between many flashes, and dividing by the number of flash periods can give a good approximation for the rotation period. However, due to the synodic effect, the flash can occur a little later or a little earlier than expected, and therefore twice the flash period does not always equal the rotational period.

For the first few times of observing flashing satellites, it is best just to follow a satellite over a few passes and become accustomed to the flash pattern and the influence of the synodic effect. When the pattern is familiar, a stopwatch can be used to time a certain number of flashes. The stopwatch should be started at a distinctive time in the flash cycle, this is usually at the flash or brightest time, but can be when it is at its dimmest or at a minimum brightness. The count should begin at zero and the stopwatch stopped on the last flash you want to observe. For example, at a count of twenty, twenty one flashes will have been seen, but only twenty flash periods. If the count had started at one, the flashes and not the periods would be counted. The total time divided by the number of periods gives the flash period. The more flash periods counted, the more accurate the result will be. If the stopwatch has a split time facility, taking a split timing somewhere in the middle of the count can be used to check the accuracy of the final reading, and also can be used as a backup if the final reading is unavailable, for example if clouds obscure the satellite.

The accuracy of the total time measurements should be estimated. This will depend on a variety of factors including the flash pattern, your reaction time and the visibility of the flashes. As mentioned earlier, not many people will have a reaction time of less than 0.1 seconds and most will be nearer 0.2 seconds. If there is a long pause between flashes, and the flash is indistinct, the accuracy may be out by as much as several seconds. More accurate or more complex observations can be made by recording your voice while counting the flashes and also recording an accurate time marker, for example, from a shortwave radio tuned to a time signal. The data can be analysed later to determine the flash period. Other observations can include estimating the brightness of the flash and the pattern of the flashes.

Amateur observers can contribute their measurements of the flash period of satellites to the Belgian Working Group for Satellites (BWGS), who coordinate such observations and collate the data. Their web site is in Dutch. Mike McCants' web site, which is in English, explains more about the rotation period or Photometric Periods of Artificial Satellites (PPAS), and also explains how to contribute data to the BWGS and the format that is required.

Brightness Measurement

The brightness of a satellite can be determined by comparison with nearby stars. The magnitude of stars can be found in a star atlas. Being able to determine the magnitude of the dimmest stars visible under the current atmospheric conditions will help distinguish which satellites will be able to be seen. By using binoculars or a telescope much fainter objects can be seen. However, the field of view is less and they therefore must be aligned more accurately. As a rough guide, with just the naked eye, the limit is about magnitude 6. A pair of 50 millimetre binoculars will enable objects of about magnitude 10 to be seen, if the skies are suitable. A 150–200 millimetre reflector telescope will increase the seeing to about magnitude 14. To follow a satellite across the sky with a telescope can be difficult, and a motor driven mount is usually required although it may not track with sufficient accuracy or be able to follow the satellite quickly enough.

Glossary

ablation — The removal of surface material by vaporization or melting, usually used to provide thermal protection.

abort — To cancel or cut short a flight.

albedo — The fraction of light or heat reflected from a body such as a planet or a moon.

Ansari X-Prize — Competition for the first non-governmental organization to launch a reusable manned spacecraft into space twice within 2 weeks. The US$10 million prize was won by SpaceShipOne in 2004.

aperture — Diameter of the main lens or mirror in a telescope, or the diameter of the collecting dish for radio astronomy.

aphelion — The place on an **orbit** around the Sun that is furthest away from the centre of the Sun.

apogee — The place on an **orbit** around the Earth that is furthest away from the centre of the Earth.

apparent magnitude — The brightness of a star as viewed from the Earth.

arcminute — A small unit of angular measurement, one sixtieth of a degree.

Aries, First Point of — The direction of the Sun from the Earth on the **Vernal Equinox**.

asteroid — A small rocky body in **orbit** around the Sun. Most are between the orbits of Mars and Jupiter.

astronomical unit (AU) — A unit of distance based on the average distanc between the Sun and the Earth. 49,600,000 kilometres.

attitude	The orientation of a spacecraft.
ballistic trajectory	The **trajectory** followed by a body acted on only by gravitational forces.
blackout	A loss of radio communication due to environmental conditions or to ionization of the atmosphere around a re-entry vehicle; Temporary lack of vision caused by a decrease of blood pressure to the head.
burnout	The end of fuel and oxidiser burning, or the time when this happens.
celestial sphere	An imaginary sphere encompassing the Earth.
chemical rocket	A rocket that requires chemical fuel for propulsion.
declination	Angular distance north or south of the celestial equator.
degrees of freedom	Possible modes of motion, with respect to a coordinate system. For a free body there are six degrees of freedom, three linear and three angular.
desorb	To remove from the surface something which was adsorbed.
DSN	The **NASA** Deep Space Network. An international network of antennas that supports interplanetary spacecraft missions and radio and radar astronomy observations. The network also supports selected Earth-orbiting missions.
eccentricity	A measure of the elliptical shape of an **orbit** compared to a perfect circle.
ecliptic	The apparent annual path of the Sun through the stars as seen from the Earth; the intersection of the plane of the Earth's **orbit** with the **celestial sphere** (therefore the Earth has an orbital **inclination** of zero).
ellipse	A closed curve, like a flattened circle.

epoch	A fixed point in time, usually given for when the positions of celestial objects are calculated. The current epoch is January 1, 2000 at 12:00, written as J2000.0.
ESA	European Space Agency. ESA is an international organization with 17 Member States.
escape velocity	The radial speed which a body must obtain to escape from the gravitational field of a planet or star.
First Point of Aries	See **Aries, First Point of**
g	An acceleration equal to the acceleration of gravity on the Earth at sea level.
geostationary orbit	A **geosynchronous orbit** at an altitude of 35,880 kilometres. A **satellite** in a geostationary orbit appears stationary above one point on the Earth.
geosynchronous	Revolving at the same rate as the Earth spins.
gravity gradient	The change in strength of gravity with the distance between two bodies, such as the Earth and the Moon.
IAU	See **International Astronomical Union**.
ICBM	See **Intercontinental Ballistic Missile**.
IIN	See **international identification number**.
inclination	The orientation of an **orbit** with respect to the equator or other reference plane.
Intercontinental Ballistic Missile	Missile with a range greater than 5,500 kilometres.
interferometer	Instrument that uses the interference of waves to make measurements in terms of the wavelength.
International Astronomical Union (IAU)	Group whose mission is to promote and safe guard the science of astronomy through international cooperation.

international designator	See **international identification number**.
international identification number (IIN)	An international naming convention for **satellites**.
International Space Station (ISS)	A research facility currently being assembled in space.
International Telecomm unication Union	An international organization established to standardize and regulate international radio and telecommunications.
Isp	See **specific impulse**.
ISS	See **International Space Station**.
ITU	See **International Telecommunication Union**.
Jet Propulsion Laboratory	**NASA** centre run by California Institute of Technology that builds and operates unmanned spacecraft.
JPL	See **Jet Propulsion Laboratory**.
Keplerian orbit	An **orbit** that is mainly determined by gravitational forces.
laser	A device used to produce coherent light from light amplification by stimulated emission of radiation.
launch pad	The platform from which a **launch vehicle** is launched.
launch period	Number of days when a rocket can be launched to reach the desired **orbit** or **trajectory**.
launch vehicle	Any device that propels and guides a spacecraft into **orbit**.
launch window	Time within a **launch period** when a rocket can be launched to reach the desired **orbit** or **trajectory**.
lift-off	Separation of the **launch vehicle** from the ground.

light year	A unit of distance based on the distance travelled by light in a year. 9,460,000,000,000 kilometres or 63,240 **Astronomical Units**.
magnitude	The measure of the brightness of a star. The higher the magnitude, the dimmer the star will appear. For bright objects, the magnitude can be negative.
maser	A device that amplifies microwave frequencies from microwave amplification by stimulated emission of radiation.
meteor	A brief streak of light made by a small body passing through the Earth's atmosphere.
meteorite	A small natural object from space that hits the surface of the Earth or other celestial body.
meteoroid	A small rocky body in **orbit** around the Sun. Meteoroids are smaller than **asteroids**.
microgravity	See **weightless**.
momentum	A measure of the quantity of motion of an object, determined by the mass of the object multiplied by its velocity.
National Aeronautics and Space Administration (NASA)	US Space Agency.
NASA	See **National Aeronautics and Space Administration**.
National Oceanic and Atmospheric Administration (NOAA)	US federal agency focused on the condition of the oceans and the atmosphere.
NOAA	**See National Oceanic and Atmospheric Administration**.
NORAD	See **North American Air Defence Command**.

North American Air Defence Command (NORAD)	Aerospace warning and aerospace control for North America.
orbit	The path of an object under the influence of a gravitational force.
orbital velocity	The speed with which a **satellite** orbits an object. The speed balances the gravitational pull of the object.
payload	The cargo carried by the **launch vehicle**.
perigee	The place on an **orbit** around the Earth that is nearest to the centre of the Earth.
perihelion	The place on an **orbit** around the Sun that is nearest to the centre of the Sun.
perturbation	A change in the **orbit** of a body, usually due to the gravitational influence of another body.
plasma	A gas of charged particles.
precession	The wobbling motion of the Earth's axis of rotation, which gradually sweeps out a conical shape. It takes the Earth 25,800 years to complete a circle.
Right Ascension	Angular distance east of the **Aries, First Point of**, measured along the celestial equator.
satellite	Small body revolving in **orbit** around a larger body. Can be natural or man-made.
sidereal period	The time taken for a **satellite** or planet to complete one **orbit** relative to the stars.
solar cycle	The variation in the amount of solar activity, such as sunspots and solar flares, over an 11-year period.
solar flare	Violent explosion in the Sun's atmosphere.
solar wind	Flow of material from the Sun.

space debris	The remnants of rockets, **satellites** and other man-made objects left in space.
specific impulse (Isp)	A measure of a propulsion system's power and efficiency at using the propellant.
sublimation	The vaporization of a solid into a gas, without an intermediate liquid stage.
supernova	A star that undergoes a sudden and temporary increase in brightness as a result of an explosion.
TDRS	See **Tracking and Data Relay Satellite System**.
thrust	The pushing force exerted on a spacecraft by an engine or motor.
Tracking and Data Relay Satellite System (TDRS)	Communication signal relay system that provides tracking and data acquisition services between spacecraft and control and/or data processing facilities.
tracking, telemetry and command (TT&C)	Subsystem on a spacecraft.
trajectory	The path of an object that is not in an **orbit**.
TT&C	See **tracking, telemetry and command**.
universal time (UT)	A standard timescale, the same as Greenwich Mean Time (GMT).
US Air Force Space Command (USSPACECOM)	Coordinated the use of US Army, Naval and Air Force space forces; Disbanded on October 1, 2002, its responsibilities were handed over to **USSTRATCOM**.
US Air Force Strategic Com mand (USSTRAT COM)	Command and control centre for US strategic forces.
USSPACECOM	See **US Air Force Space Command**.
USSTRATCOM	See **US Air Force Strategic Command**.

Vernal Equinox	When the Sun passes from south to north across the celestial equator. The length of daylight and darkness is exactly equal.
weightless	Condition in which no acceleration can be felt, either from gravity or from another force.
zero G	See **weightless**.

Bibliography

Bate, Roger R., Mueller, Donald D., White, Jerry E., *Fundamentals of Astrodynamics*, Dover, 1971, ISBN 9-780486-6006-11
Bergerac, Cyrano de, *Histoire Comique des Etats et Empires de la Lune et du Soleil*, 1657
Chaikin, A., *A Man on the Moon*, Penguin, 1998, ISBN 0-14-024146-9
Clarke, Sir Arthur C., *Extra-Terrestrial Relays*, 1945
Clarke, Sir Arthur C., *The Fountains of Paradise*, Victor Gollancz, 1979, ISBN 0-575-02520-4
Davidoff, M., *The Radio Amateur's Satellite Handbook*, ARRL, 2003, ISBN 0-87259-658-3
Dornberger, Major-General W., Cleugh, J., Halliday, G., *V2*, Hurst & Blackett, 1954
Fortescue, P., Stark, J., Swinerd, G., *Spacecraft Systems Engineering*, Wiley, 2003, ISBN 13:978-0-471-61951-2
Gilster, P., *Centauri Dreams*, Copernicus Books, 2004, ISBN 0-387-00436-X
Goddard, Robert H., *A Method of Reaching Extreme Altitudes*, 1919
Godwin, F., the Bishop of Hereford, *A Man in the Moon*, 1638
Gunston, B., *The Cambridge Aerospace Dictionary*, Cambridge University Press, 2004, ISBN 0*521-84140-2
Harris, Philip R., *Living and Working in Space*, Wiley, 1996, ISBN 0-471-96256-2
Kepler, J., *The Somnium*, 1630
King-Hele, D., *Observing Earth Satellites*, Macmillan, London, 1983, ISBN 0-333-33041-2
Larson, Wiley J., Wertz, James R., *Space Mission Analysis and Design*, Space Technology Library, 1999, ISBN 1-881883-10-8
Lovell, B., *The Story of Jodrell Bank*, 1968
Lucian of Samosata, *True History or True Story*, 160 AD
McInnes, Colin R., *Solar Sailing*, Springer-Praxis, 1999, ISBN 3-540-21062-8
Miles, H., *Artificial Satellite Observing*, 1974, ISBN 0-571-09600-X
NASA, *NASA Thesaurus*
Chartrand, Mark R., *Collins Guide to the Night Sky*, 1999, ISBN 0-00-2200147
Nye, N., *Art of Gunnery*, 1647
Oberth, H., *Die Rakete zu den Planetenräumen*, 1923
Oberth, H., *Wege zur Raumschiffahrt*, 1929
Ordway, Frederick I., Gardner, James P., Shrape, Mitchell R., *Applied Astronautics*, Prentice-Hall, 1963, Library of Congress Catalogue Number 63–9521
Ordway, Frederick I., Gardner, James P., Shrape, Mitchell R., *Basic Astronautics*, Prentice-Hall, 1962, Library of Congress Catalogue Number 05677-C
Ridpath, I., *Norton's Star Atlas and Reference Book*, Pi Press, 2004, ISBN 0-13-145164-2
Ridpath, I., *Oxford Dictionary of Astronomy*, Oxford University Press, 1997, ISBN 0-19-211596-0
Rosen, E., *Kepler's Somnium*, Dover, 2003, ISBN 0-486-43282-3
Siemienowicz, K., *Artis Magnae Artilleriae pars prima*, 1650
Smith, S.W., *A Handbook of Astronautics*, 1966

Sutton, George P., Biblarz, O., *Rocket Propulsion Elements*, Wiley, 2001, ISBN 0-471-32642-9
Tsiolkovsky, K., *The Exploration of Cosmic Space by Means of Reaction Devices*, 1903
Various Space Agencies, *The Global Exploration Strategy: The Framework for Coordination*, 2007
Verne, J., *Autour de la lune*, 1870
Verne, J., *De la terre à la lune*, 1865
Wilkins, Reverend Dr. J., *A Discourse Concerning a New Planet*, 1640
Wilkins, Reverend Dr. J., *Mathematical Magic*, 1648
Wilkins, Reverend Dr. J., *The Discovery of a World in the Moon*, 1638
Willis, Edward A., *Manned Venus Orbiting Mission*, NASA Technical Memorandum, Document Number TM X-52311, 1967

Index

References to figures are printed in bold.

Printed in Great Britain
by Amazon